General Editor's Preface

Defects in CrystallineSolids is the first title to appear in a series of co-ordinated texts on Solid State Physics. The aim of the series is to provide a coverage of those aspects of the solid state which can appropriately be considered in undergraduate physics or materials science courses as well as in post-graduate courses in a number of other departments. The series will offer a selection of topics covering all those aspects which form the core of a subject which has grown very rapidly in the last twenty years.

Other titles will include *The Crystal Structures of Solids*, *The Dynamics of Atoms in Crystals*, *The Electrical Properties of Solids*, *The Electronic Structures of Solids* and *The Magnetic Properties of Solids*. From a traditional viewpoint that starts with the simplest possible solid *Defects in Crystalline Solids* might seem to come fairly late, but to the science student led to consider solids because of their more interesting or technologically important properties, it lies at the very gateway of the subject and it has the advantage of being suitable for a self contained treatment. Dr. Henderson has made important research contributions in this field and has long experience of presenting it to students; I hope the combination of authoritative treatment and characteristic flavour which he provides will come to be regarded as the hallmark of all the books in this series.

Imperial College,
London,
1972.

B.R.C.

Preface

The purpose of this text is to present some of the basic physical phenomena necessary for the understanding of the defect solid state. In part the subject matter is based upon an optional course in Physics which I taught to Final year undergraduates at the University of Keele, England. The book is, however, intended for a wider readership, including undergraduates and junior postgraduates of Physics, Chemistry, Metallurgy and Materials Science.

Over several decades, the study of crystalline lattice defects has derived a considerable impetus from the need to understand the control exerted by defects on most technologically important properties of solids. When used in the most general context, the term *defect* might encompass such elementary excitations as phonons, excitons, polarons, magnons and plasmons, in addition to the many static structural imperfections. Such a broad interpretation is beyond the scope of the present discussion: instead I have attempted to outline some of the experimental phenomena associated with the presence of holes and electrons, impurity atoms, vacant and interstitial lattice sites, as well as line and planar defects in a variety of crystalline solids. The first two chapters are concerned with the specification and energies of defect types, and the experimental techniques used in investigating lattice defects. The following chapters are concerned with point defects in ionic solids, semi-conductors and metals. The final chapter is a short account of some consequences of interactions between defects in metals. In such a short account of the factors controlling the mechanical properties of solids it is only possible to give a flavour of the present situation. Even so discussion of the theory of work hardening has been completely omitted. The book is strongly biased towards point defects and small clusters of point defects. This represents a major difference between the present text and other treatises on solid state physics and materials science, which usually give a major coverage to dislocation theory and its application to the mechanical properties of solids. It is now appropriate, I hope, to attempt to redress the balance.

The book has an extensive but by no means comprehensive bibliography; where a reference is appended to an illustration it has not usually been duplicated in the text. In addition, a few general references are given at the

The Structures and Properties of Solids 1

Defects in Crystalline Solids

B. Henderson, B.Sc, Ph.D, F.Inst.P
Reader in Physics, Keele University

Edward Arnold

First published 1972 by Edward Arnold (Publishers) Ltd.
25 Hill Street, London W1X 8LL

Boards Edn. ISBN: 0 7131 2350 8
Limp Edn. ISBN: 0 7131 2351 6

Set at The Universities Press, Belfast, and
printed in Great Britain by The Pitman Press, Bath.

end of some chapters. Readers seeking to remind themselves of the descriptions of perfect crystal structures given by formal crystallography and drawn upon at some points are referred to the companion book in this series *The Crystal Structures of Solids* by Forsyth and Brown.

It is a pleasure to record my indebtedness to various friends and colleagues who have helped during the preparation of this volume; to Professor R. E. Smallman who first stimulated my interest in defects, to present colleagues at Keele and to past colleagues, especially Dr. F. J. P. Clarke and the Basic Ceramics Group at Harwell, for their encouragement and stimulating company, and to Professor B. R. Coles at whose suggestion the book was written. In addition I am particularly grateful to the numerous authors who have supplied original negatives for the reproduction of the various micrographs. I am also grateful to all authors and publishers who have given permission to use their original material. To my wife and to Mrs. B. Ford I owe a particular debt of gratitude for patiently deciphering the hieroglyphics I pass for handwriting and for producing the final manuscript.

1972

B.H.

Contents

1

Defects in Solids

For several decades theoretical physicists have attempted to deduce the properties of solids from the properties of their constituent atoms. It is notable that although almost all solid state phenomena can now be explained qualitatively, the modern theory of solids is not yet able to predict in advance the properties of any given aggregate of atoms. This is not surprising since from a completely satisfactory theory one should be able to deduce the crystal form as well as the mechanical, thermal, electronic and magnetic properties from the properties of atoms; a many-body problem of daunting complexity. Even calculations of cohesive energy are not yet sufficiently precise to discriminate for any one solid between the various crystal structures possible. Fortunately, the atomic arrangement in many solids is regular and highly ordered which greatly simplifies the calculative complexity.

The internal, long-range regularity of atomic packing is frequently reflected in the external form of synthetic and naturally occurring crystals. In these solids quite beautiful confirmation and extension of the evidence for crystallinity is obtained from x-ray diffraction studies. However, analysis of such studies frequently implies a perfection of crystal structure that is not apparent in *real* solids. For example, the atomic vibrations are structural imperfections since they cause atoms to be displaced from the equilibrium lattice positions assumed for the perfect crystal. Furthermore simple thermodynamic reasoning demonstrates that at equilibrium, there are present at any non-zero temperature both vacant lattice sites and interstitial atoms. Important among other defects present in real solids are:

a) substitutional and interstitial impurity atoms
b) dislocation lines
c) stacking faults and other planar defects and
d) grain boundaries.

Supplementary to these *static* imperfections are a number of elementary excitations which have sometimes been classified as lattice imperfections. These include excitons and phonons, polarons, and quantized spin waves or magnons. Except when their interaction with one of the simple static defects becomes important, these elementary excitations are not discussed further in this text.

1.1 The specification of lattice defects

The simplest structural imperfections in solids are those involving single lattice points. Impurity atoms substituted either intentionally or unintentionally constitute such *point defects*. The difference in size between an impurity atom and the solvent atoms produces a centre of elastic distortion in the lattice. Provided that this elastic distortion is small, the increase ΔE in the crystal's internal energy E is also small. The change in configurational entropy ΔS, which is related to the number of ways of distributing a small number of impurity atoms among a large number of lattice sites, is however quite large. Thus at finite temperature T, the presence of impurity atoms in the crystal is thermodynamically desirable, since the free energy $F = E - ST$ may be reduced by their presence. This result has important implications in respect of the ultimate purity attainable in elemental solids.

Impurity atoms may also occupy interstitial sites, i.e., they may be squeezed into positions between the solvent atoms. This may result in a large increase in both the internal energy and the configurational entropy: consequently interstitial solid solutions are rare since they are confined to very small atoms being present in the interstices of relatively open structures. The most important technological example of an interstitial solid solution is

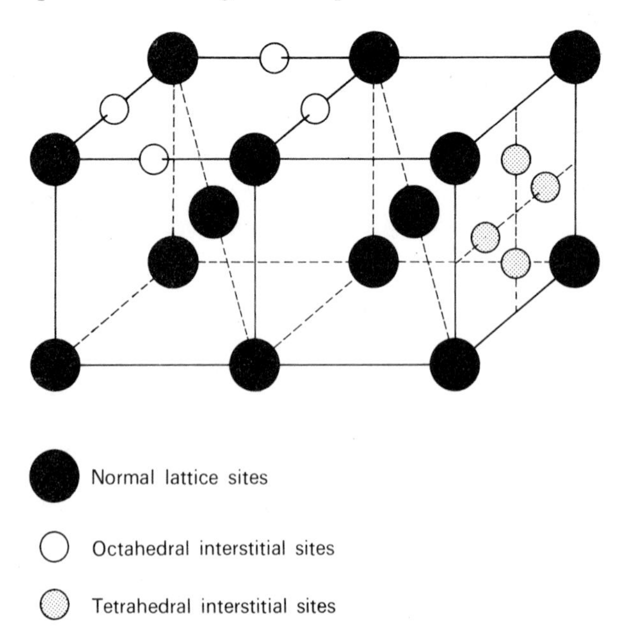

Fig. 1.1 Tetrahedral and octahedral interstices in a body-centred cube. The largest interstitial site occurs at ($\frac{1}{2}$ $\frac{1}{4}$ 0) and equivalent positions in the unit cell: it has tetrahedral symmetry. The octahedral site at the mid-point of the cell edges, e.g. at (0 0 $\frac{1}{2}$), is smaller but is preferentially occupied by carbon and nitrogen atoms in α-iron.

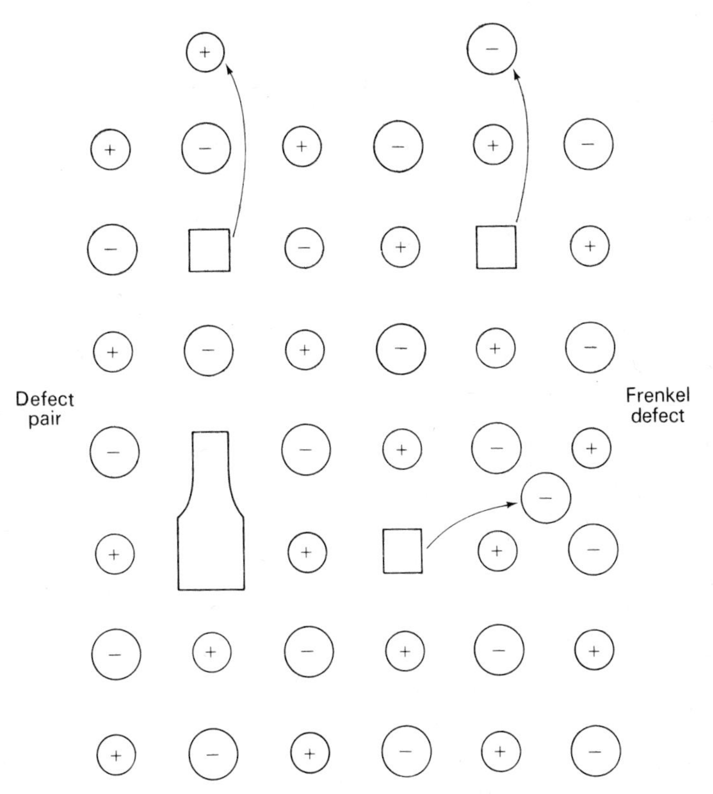

Fig. 1.2 Point defects in the rock-salt structure, showing a Schottky defect, a Frenkel defect and a vacancy pair.

the solution of carbon or nitrogen in iron. At elevated temperatures iron exists as a face-centred cubic solid, γ-iron, in which the largest interstice, approximately 0·52 Å in radius, is at the centre of the unit cell. Despite having an atomic radius of 0·7 Å, carbon dissolves interstitially in γ-iron to the extent of almost 8 atomic per cent. There is a concomitant increase in the size of the unit cell. In body-centred cubic iron, α-iron, the largest interstitial site is a tetrahedral site, four-fold co-ordinated with four iron atoms situated symmetrically around it: this site can accept an atom of radius 0·36 Å. Surprisingly the carbon atoms prefer to occupy an even smaller interstitial site. This preferred site has distorted octahedral symmetry and can only accommodate an atom of radius 0·19 Å. The two sites are shown in Fig. 1.1: the preference for the octahedral site is apparently related to the anisotropic elastic constants of the body-centred cubic lattice.

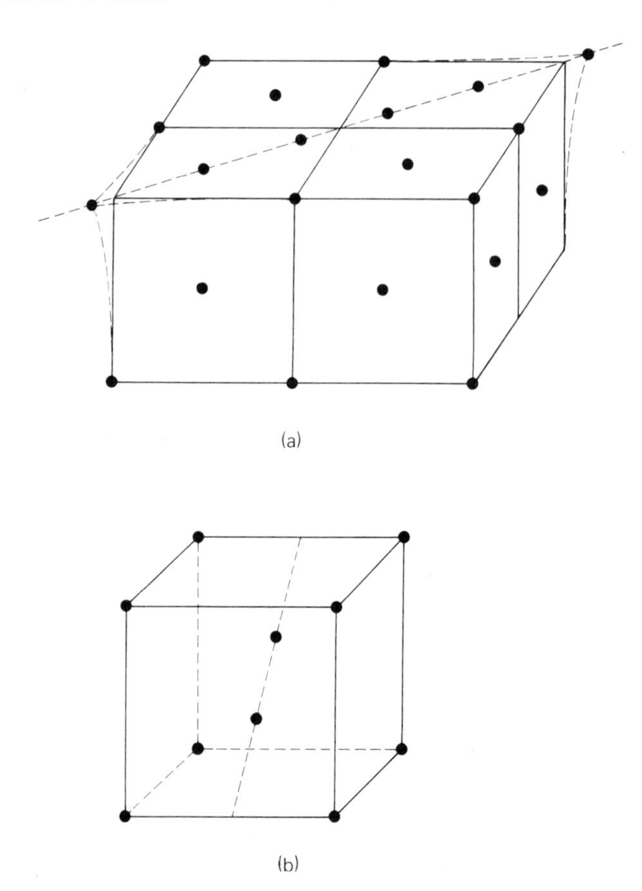

Fig. 1.3 Complex defect structures in cubic lattices. In (a) the "crowdion" emphasizes that several atoms are displaced from their normal sites to accommodate an interstitial atom in a face-centered cube. (b) shows a "split-interstitial" in a body-centred cube.

Arguments similar to those given above suggest that both vacancies and interstitial atoms (not impurities) exist in thermal equilibrium in all crystals. These intrinsic lattice defects are produced when an atom is removed from a normal lattice site to some other place in the crystal. If the atom moves to a crystal surface or to some other surface within the body of the crystal, the resulting vacancy is called a *Schottky* defect. Another vacancy defect is the *Frenkel* defect which results when the displaced atom is lodged in an interstitial site in the lattice. It is obvious that the defect with the lowest energy of formation will predominate. Experimental evidence suggests that in both metals and the alkali halides the most common defects are Schottky defects. In pure silver halides Frenkel defects predominate. In ionic solids such as the alkali and silver halides, the lattice defects must be produced in such a way as to maintain both stoichiometry and charge neutrality. This is shown schematically in Fig. 1.2: the Schottky defect is a pair of vacancies, one on each of the anion and cation sub-lattices. The Frenkel defect, however, is still a single vacancy and a single interstitial, since whichever sub-lattice it occurs on, the vacant lattice site is oppositely charged to the interstitial ion, on account of the net charge associated with lattice ions surrounding the vacancy. A further consequence of the ionic structure of the alkali halides is that an electrostatic attraction exists between anion and cation vacancies. Thus there is a tendency for such vacancies to form neutral vacancy pairs in concentrations related to the binding energy.

The simple defect configurations indicated in Fig. 1.1 and 1.2 are not necessarily the most stable structures. For example, a pair of closely associated vacancies might have a lower energy than the two defects considered separately. Similarly an interstitial atom in a line of close packed atoms may

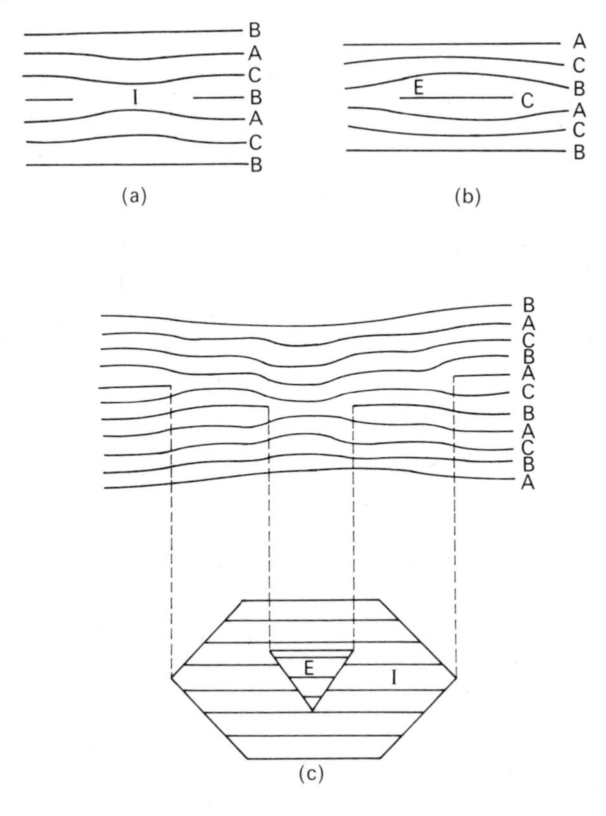

Fig. 1.4 The formation of dislocation loops by aggregation of (a) interstitials and (b) vacancies. The structure of a double faulted loop is shown in (c).

be thought of as being shared by several atoms in the row (Fig. 1.3a). Such a "crowdion" configuration has the interesting property of requiring only a small activation energy for propagation along the row of atoms. Such a mechanism might be important in low energy diffusion processes, especially in irradiated solids. Figure 1.3b shows that the "split-interstitial" is similar to the crowdion, since two atoms are permitted to share one lattice site. Very detailed computer calculations have lead to a number of interesting configurations for point defect complexes involving three and four vacancies. According to such calculations the most stable configuration for the tri-vacancy in copper has tetrahedral symmetry involving four vacant lattice sites with one atom relaxed into the centre of the cavity. The quadravacancy involves six adjacent vacancies with two atoms relaxed into the centre of the multiple-vacancy. There is little direct experimental evidence for the existence of multiple vacancies and interstitials in metals. In the alkali halides and in the alkaline earth oxides the evidence for vacancy-pairs and larger aggregates is much more compelling.

Under certain conditions (e.g., after quenching, plastic deformation or irradiation) a large supersaturation of point imperfections may exist: such conditions may lead to a more extensive form of defect aggregation. The very strong yield point phenomenon in mild steel is concerned with the clustering of carbon atoms around *dislocation lines* (Chapter 7). Dislocation lines may be produced under extreme conditions when point defects coagulate on some atomic plane to form a sheet or disc of point defects as shown in Fig. 1.4. Evidently if the aggregate of vacancies is sufficiently large the adjoining planes will relax inwards with the result that the entire sheet of vacancies is bounded by a *dislocation loop*. Precipitation of vacancies on the close-packed planes of a face-centred cubic structure will produce an *intrinsic fault* (*I* in Fig. 1.4a) in which the stacking sequence above and below the vacancy loop is continuous right up to the loop. The condensation of interstitial atoms on close-packed planes gives rise to an *extrinsic fault*. In an extrinsic fault, shown as *E* in Fig. 1.4b, there are two breaks in the stacking sequence, one above and one below the fault plane. Thus an extrinsic fault will have a larger characteristic energy per unit area than an intrinsic fault.

Transmission electron micrographs which provide evidence of vacancy condensation in aluminium alloys are shown in Figs. 1.5a and b. The faults are seen to have regular geometric shapes and occasionally to show fringed contrast. The images with and without contrast in Fig. 1.5a are from different types of fault and must not be confused. For example, intrinsic faults such as that shown schematically in Fig. 1.4a would give rise to interference fringes inside the dislocation loop when observed in the electron microscope. However, in materials where the stacking fault energy is high, the intrinsic fault may be unstable. Consequently, subsequent structural changes can occur which eliminate the fault whilst increasing the distortion around the dislocation. In Fig. 1.5a, which shows dislocation loops in a

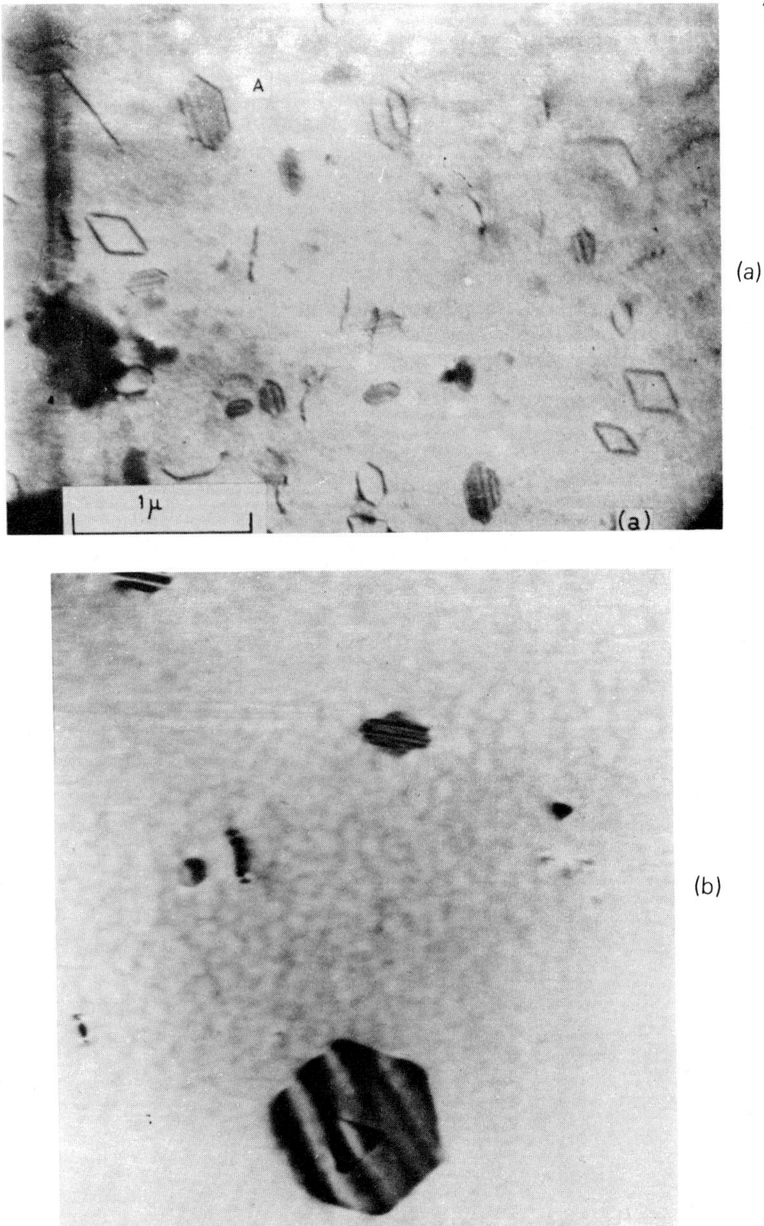

Fig. 1.5 Electron micrographs of faults in metallic crystals showing (a) Frank sessile loops and prismatic dislocation loops in aluminium (b) a double loop in an Al-0·65 wt.%Mg alloy, (c) and (d) stacking fault movement in stainless steel. (Reprinted by kind permission of Professor R. E. Smallman,(a) and (b). Reprinted from Whelan M. J., Hirsch P. B., Horne R. W., Bollman W., *Proc. Roy. Soc.*, **260,** 1957**A.** 530,(c) and (d).)

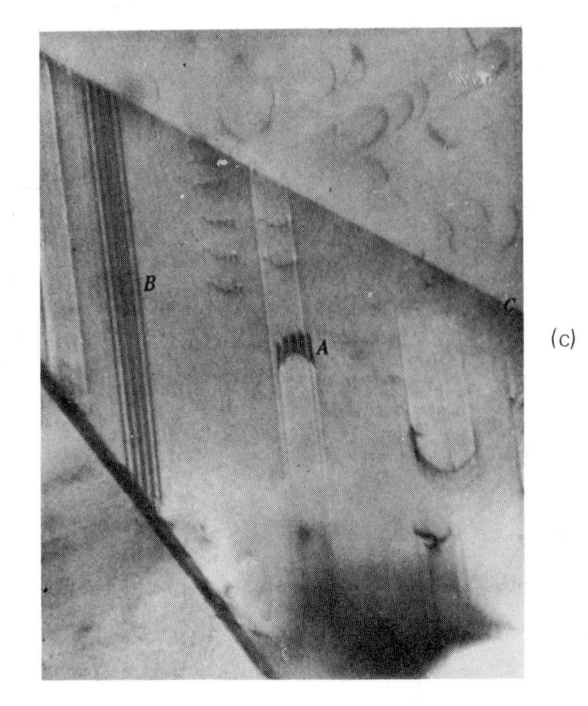

(c)

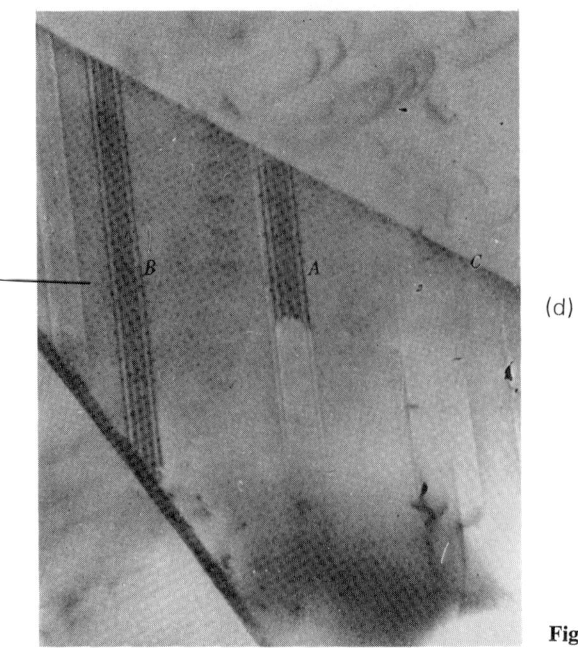

(d)

Fig. 1.5 (c) and (d)

quenched aluminium $-3{\cdot}5\%$ Mg alloy, it is clear that no stacking fault contrast is observed inside the vacancy in most cases. Thus Al is likely to have a relatively high stacking fault energy.

More complex loops formed by multiple layers of vacancies being precipitated on top of one another have also been observed in several quenched metals (e.g., Al, Zn and Mg). The addition of each layer of vacancies alternately introduces or removes stacking fault. Figure 1.4c shows a schematic representation of a double loop from which it is evident that the inner region is extrinsically faulted and the outer region is intrinsically faulted. The appearance of such a doubly faulted loop in an aluminium alloy is shown in Fig. 1.5b. Although the inner region of a double fault is usually triangular in shape with sides parallel to three of the hexagon sides it is not always so. On annealing such a large double loop, vacancies are emitted from the dislocation line of the hexagon which shrinks and becomes circular. At the same time the inner loop grows by vacancy absorption. From a detailed study of the kinetics of this process it has been shown that the ratio of the extrinsic/intrinsic fault energy is $\simeq 1{\cdot}4$ for aluminium (see also §6.2).

1.2 Point defects in thermal equilibrium

At a finite temperature T, the equilibrium state of a solid is determined by minimizing the Gibbs free energy $F = E - TS$: we saw above that this condition implies the existence of some lattice disorder above the absolute zero in temperature. Both the energy required to produce defects in the static lattice and the changed vibrational energy of the atoms neighbouring the defects contribute to the change in internal energy consequent upon the introduction of defects into the crystal. There are also two contributions to the change in entropy experienced by the crystal. Firstly the thermal entropy changes as a result of the frequency change of the atomic oscillators, and secondly there is a configurational entropy term associated with the number of ways of selecting which atomic sites will be vacant. In the following calculation we consider only the formation of Schottky defects in a monatomic lattice with cubic symmetry. The method is, however, perfectly general and is readily adapted to other systems.

Consider that there are n Schottky vacancies and $N + n$ lattice sites in the crystal including the vacancies. The energy of formation E_S of each imperfection is clearly the energy, E_H, required to take an atom to the crystal surface from the interior minus the lattice energy per atom E_L. Thus the internal energy of the crystal is changed by an amount

$$nE_S = n(E_H - E_L)$$

and the internal energy of the disordered crystal E is

$$E = E_P + nE_S$$

E_P referring to the internal energy of the perfect lattice. We calculate the thermal entropy change of the crystal using the Einstein model of the vibrating lattice, in which each atom vibrates with frequency ν independently of all other atoms. Neglecting the zero point vibrational energy the thermal entropy[1] is,

$$S = 3Nk \ln (h\nu/kT)$$

In the disordered lattice, those atoms next to a vacancy have a smaller vibrational frequency ν' along the line joining the atom to the vacancy than the perfect lattice frequency, since the restoring forces are reduced. Thus we assume that each atom neighbouring a vacancy is equivalent to three independent harmonic oscillators, two with frequency ν and one with frequency ν'. Consequently in a lattice site with x-fold co-ordination, there are nx atoms oscillating with frequency ν' and $(3N - nx)$ atoms oscillating with frequency ν. The configurational entropy term is obtained from the Boltzmann equation,

$$s = k \ln W$$

where

$$W = \frac{(N+n)!}{N!\,n!}$$

and represents the number of ways of arranging the vacancies in the lattice.

Hence the free energy of the disordered crystal is given by,

$$F = E_P + nE_S - (3N - nx)kT \ln\left(\frac{h\nu}{kT}\right) - nxkT \ln\left(\frac{h\nu'}{kT}\right) - kT \ln \frac{(N+n)!}{n!\,N!}$$

The equilibrium value of n is obtained from this equation by setting $(\partial F/\partial n)|_T = 0$, and using Stirling's approximation,

$$\ln N! = N \ln N$$

for expanding the logarithms of large numbers. Hence

$$\left.\frac{\partial F}{\partial n}\right|_T = E_S - xkT \ln\left(\frac{\nu'}{\nu}\right) - kT \ln\left(\frac{n}{N+n}\right) = 0$$

and consequently

$$\boxed{\frac{n}{N} = \left(\frac{\nu}{\nu'}\right)^x \exp - \left(\frac{E_S}{kT}\right)} \qquad (1.1)$$

when $n \ll N$. Thus measurements of n/N as a function of temperature actually measure E_S. Applying similar arguments to the thermal equilibrium of Frenkel defects gives

$$\frac{n}{(NN_i)^{\frac{1}{2}}} = \gamma \exp - \left(\frac{E_F}{2kT}\right) \tag{1.2}$$

in which E_F is the energy to form one Frenkel defect, and N_i is the number of interstitial sites in the lattice. The pre-exponential term γ is related to the thermal entropy changes through the changes in the vibrational frequencies of those atoms neighbouring both the interstitial and the vacancy. That part of γ related to the vacancy component of the Frenkel defect will be $(\nu/\nu')^x$ as in Equ. 1.1. If it is assumed that the interstitial site has y-fold co-ordination and that the frequencies of vibration of the interstitial and its immediate neighbours are ν_i and ν_i respectively, it is easily shown that,

$$\gamma^2 = \left(\frac{\nu}{\nu'}\right)^x \frac{\nu^{y+1}}{\nu_i \nu_i'^y} \tag{1.3}$$

This result has the following general implication: since $\nu > \nu'$ the change in thermal entropy favours the formation of Schottky vacancies. For example with $\nu = 2\nu'$ in a cubic structure ($x = 6$) we find that $\gamma = 64$. However when Frenkel defects are formed ν_i and ν_i' are almost certainly greater than ν so that γ may be less than unity. Consequently the changes in thermal entropy do not necessarily favour the formation of Frenkel defects. The factor 2 which occurs in the exponential term in Equ. 1.2 arises because the Frenkel defect has two components. This is a quite general result and applies in the case of the Schottky defect in the alkali halides, where a cation + anion vacancy is formed simultaneously to maintain thermal equilibrium.

The preceding arguments show that extensive lattice disorder persists in equilibrium only at high temperature at which defects are expected to be mobile. Thus both the formation and mobility of atomic defects are important processes in the diffusive motion of atoms in solids. Interstitial diffusion of carbon atoms in steel occurs when the carbon atoms jump from one interstitial site to another by squeezing between the atoms around those sites. Since work has to be done against the elastic forces of the crystal the interstitial atom must acquire sufficient energy to surmount the energy "barrier" opposing the jump. The probability of such an event occurring during one of the interstitial's vibrations is obtained from the Maxwell-Boltzmann relationship;

$$P_i = A\nu_i \exp - \left(\frac{E_i}{kT}\right)$$

where E_i is the energy for movement of an interstitial and A is related to the entropy difference between the atom prior to jumping and mid-way between its initial and final positions. The total probability P is obtained by a summation over all the positions into which the interstitial may jump. The diffusion coefficient D is related to P through the equation,

$$D = fP\lambda^2 \tag{1.4}$$

where λ is the distance travelled in a single jump, and f measures the probability of the atom jumping in one direction relative to any other equivalent direction. Of the interstitial positions in body-centred cubic iron, only two-thirds of the positions allow both forward and backward movements (Fig. 1.1). Jumps perpendicular to this direction are also permitted, and the relative probability of a forward movement is therefore $\frac{1}{4}$. There are, however, four equivalent jumps for any of the interstitials, hence $f = \frac{2}{3}$. Thus with $\lambda = a/2$ where a is the lattice spacing we obtain,

$$D = \frac{a^2 \nu_i A}{6} \exp - \left(\frac{E_i}{kT}\right)$$

or

$$\boxed{D = D_o \exp - \left(\frac{E_i}{kT}\right)} \tag{1.5}$$

where $D_o = a^2\nu_i A/6$. Using the known values of D_o, a and ν_i for carbon in body-centred cubic iron we find $A \simeq 10$.

In pure solids, self-diffusion occurs mainly through the continual movement of thermally-created vacancies. Contributions to the activation energy for vacancy migration come from the diffusing atom's need to overcome the binding forces of its own site, and from the energy required by the diffusing atom to squeeze between its neighbours into the vacant site. This vacancy diffusion mechanism is also important in the diffusion of substitutional impurities in solids. The probability that any atom may jump into the next site depends upon the vacancy concentration (i.e., the probability that the next site is vacant) and upon the probability that it may acquire sufficient energy to make the jump. Consequently we may write the self-diffusion coefficient in the crystal as

$$D = D_o \exp - \left(\frac{E_S}{kT}\right) \exp - \left(\frac{E_M}{kT}\right)$$

in which D_o is related to the vibrational frequencies of the lattice and to the thermal entropy changes consequent upon the vacancies being formed and diffusing, and E_M is the vacancy migration energy. We can now define the activation energy for self-diffusion as

$$E_D = E_S + E_M$$

since

$$D = D_o \exp - \frac{(E_S + E_M)}{kT} = D_o \exp - \left(\frac{E_D}{kT}\right) \tag{1.6}$$

It is emphasized that the constant f inherently present in D_o is a geometric factor, which varies from one crystal structure to another. Self-diffusion is usually investigated by making use of radioactive tracer techniques: such measurements yield values for D_o in the range 10^{-1} to 10^2 cm^2 sec^{-1} and $E_D \approx 0{\cdot}5$ to 3 eV per atom. A few examples of the commercially important processes influenced by atomic diffusion are heat treatment, age hardening of alloys, surface hardening of steel, sintering, creep and corrosion of metals and alloys.

1.3 Some properties of dislocations

The nature of slip in crystals

In §1.1 we discussed briefly the formation of dislocation loops consequent upon vacancy or interstitial clustering. The dislocation is, however, a much more general solid state phenomenon, which was proposed to account for the discrepancy between the theoretical and experimental strength of solids. Simple theoretical estimates suggest that *perfect* crystals can withstand maximum shear stresses of 0·01 to 0·02 μ, μ being the elastic modulus.[2] This represents the stress required to cause two atomic planes to *slip* past each other in a homogeneously strained crystal. Since μ is usually of the order of 10^{12} dyn cm^{-2}, stresses near 10^{10} dyn cm^{-2} should be required to initiate slip in a perfect crystal. Such high strengths are rarely attained in real crystals, where the maximum shear stress is usually limited to 10^{-5} μ by the presence of either dislocations in ductile crystals, or micro-cracks in brittle materials. However, in very fine fibres or single crystal whiskers, stresses approaching the ideal shear strength have been observed. This has lead to the development of *fibre reinforced* solids for engineering purposes.

Figure 1.6 illustrates with a perfect two-dimensional model of a crystal, the essential ingredients of the process of plastic deformation in a crystalline solid. According to this simple model, the atoms above the slip plane must be simultaneously lifted atop the atoms in neighbouring positions below the slip plane, before they can slide down into their new positions. The periodic nature of the interatomic forces of the crystal requires the atoms to slip from one equilibrium position to another. Hence the process of slip must always connect one equilibrium position with another, such that its magnitude is an integral number of primitive translation vectors of the lattice. Although this may occur in any direction and on any plane in the crystal,

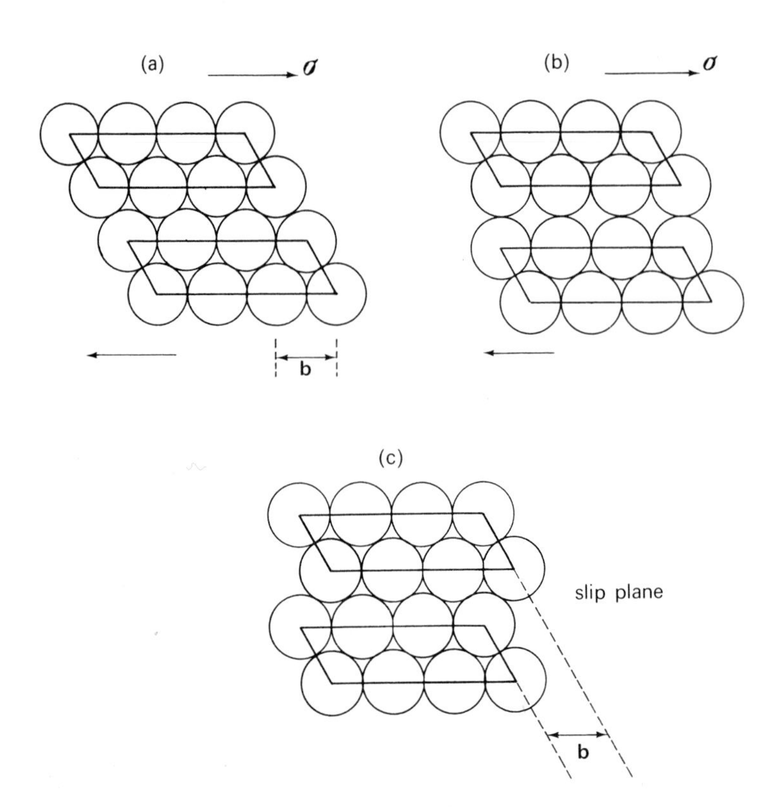

Fig. 1.6 Representing slip in a perfect 2-dimensional close-packed crystal under an applied shear stress σ. In (i) no slip has occurred, (ii) an intermediate position of the atomic planes while slip is taking place, (iii) the slip process is complete, the atoms above the slip plane having moved by one unit lattice vector, ***b***, relative to atoms below the slip plane.

slip usually occurs only in certain preferred orientations in closepacked structures. As a general rule, slip takes place only in the direction and on the plane of densest atomic packing. Thus face-centred cubic metals slip in the $\langle 110 \rangle$ directions and on $\{111\}$ planes. Body-centred cubic metals have $\langle 111 \rangle$ $\{112\}$ slip systems, although at high temperatures slip on the $\{110\}$ and $\{123\}$ planes may become operative. The preferred slip systems in close-packed hexagonal structures depend critically upon the magnitude of the axial ratio, although the close-packed $\langle 11\bar{2}0 \rangle$ directions are always involved. In magnesium, where the axial ratio is almost ideal ($c/a = 1{\cdot}633$), low temperature slip occurs on the (0001) basal planes. Above 230°C slip on the $\{10\bar{1}1\}$ planes has been observed. When $c/a > 1{\cdot}633$ the (0001) basal planes have the greatest atomic density, and slip occurs preferentially on these planes. In metals such as titanium and zirconium, where the axial ratio is appreciably less than ideal, the operative slip systems are $\langle 1120 \rangle$ $\{1010\}$ at low temperatures, and $\langle 1120 \rangle$ $\{1011\}$ at elevated temperatures.

The inadequacy of the simple theories of the ultimate strength of crystals is due to their requirement that the atomic planes as rigid entities move over one another simultaneously. Thermal vibrations of the lattice may actually help to start slip, since they introduce stress fluctuations which temporarily allow the maximum shear stress to be exceeded in a small critical volume of the crystal. Thus slip is envisaged as a stress-induced process of nucleation and growth, which starts in a localized region of the lattice and then spreads out over the remainder of the plane. The boundary between slipped and unslipped regions of the lattice is the *dislocation line*. An applied shear stress causes the dislocation line to move on the slip plane, thereby extending the slipped region of the lattice. Thus the lattice near the dislocation must be in a state of metastable equilibrium, the atoms there having higher energies than those atoms in equilibrium positions in the slipped and unslipped regions of the lattice. A simple metastable configuration is shown in Fig. 1.7: an extra-half row of atoms has been squeezed between two other rows above the slip plane. Although there is a severe local distortion, there is still matching of the lattice planes on either side of the dislocation. Slip

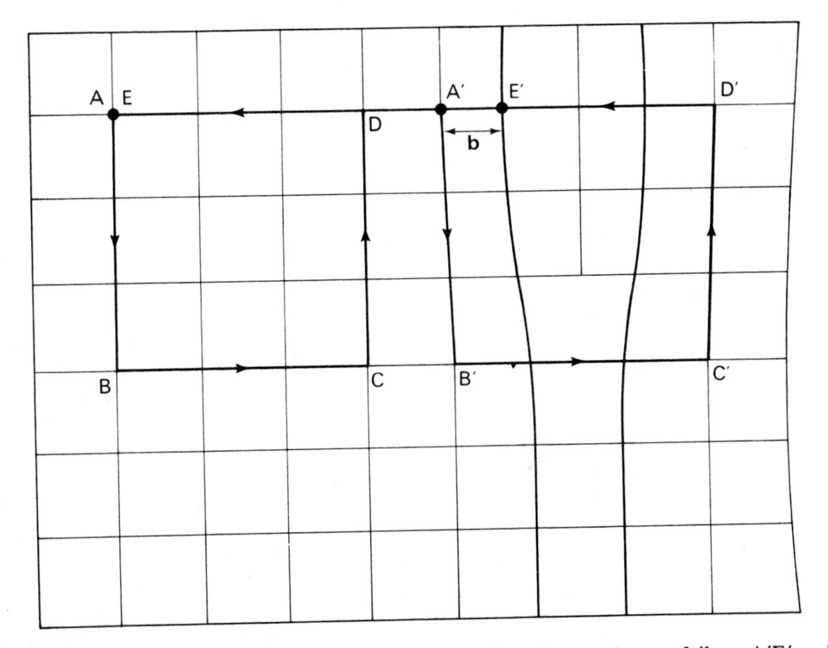

Fig. 1.7 The Burgers vector of a dislocation represented as a closure failure A′E′ = ***b*** in the circuit A′B′C′D′E′ around the dislocation, relative to a similar circuit in the dislocation free crystal.

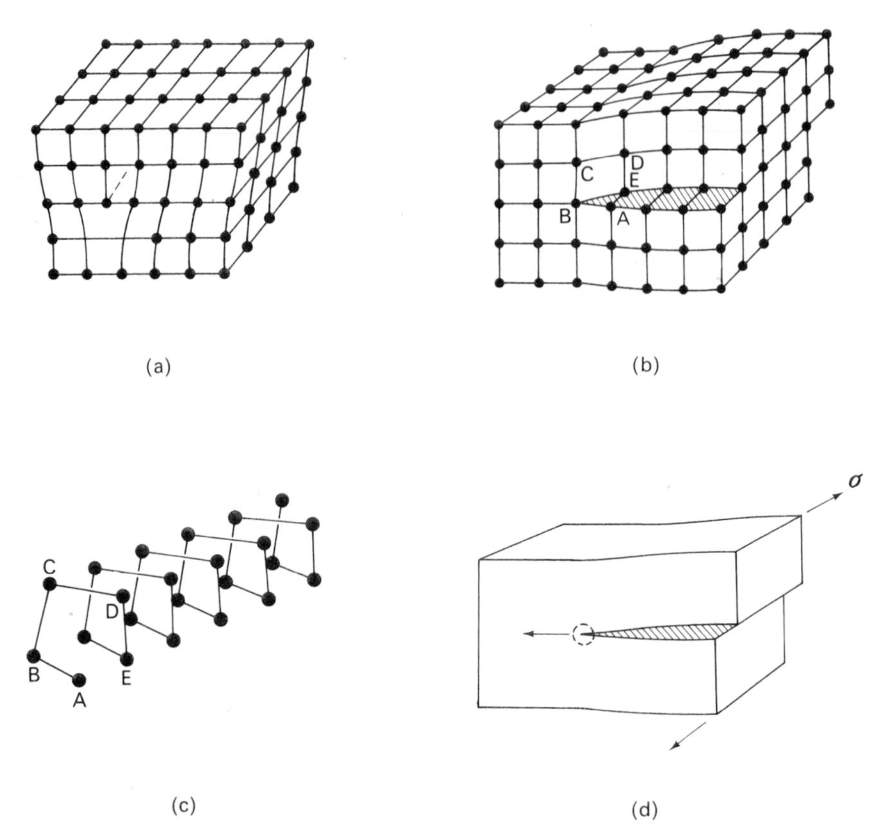

Fig. 1.8 Models of dislocations in a cubic crystal: (a) a positive edge dislocation formed by inserting an extra-half plane of atoms between atomic planes above the slip plane, (b) a right-handed screw dislocation formed by displacement of a part of the crystal above the slip-plane parallel to the dislocation line, (c) shows the spiral of atoms around the screw dislocation line and (d) the movement of the screw dislocation on the slip plane. (After HUME-ROTHERY and RAYNOR. 1960. *The Structure of Metals and Alloys.* Institute of Metals, London.)

may now propagate by the successive horizontal movements of this extra half-row of atoms. The magnitude of the slip process is defined by the Burgers vector[3], the nature of which is indicated by an atom by atom circuit along the path ABCDE in the crystal represented in Fig. 1.7. Such a circuit forms a closed loop in the perfect dislocation-free crystal. A similar circuit A′B′C′D′E′ in a crystal containing a dislocation fails to close by a single unit vector of the lattice at right angles to the dislocation line. Thus the Burgers vector is a lattice vector required to obtain a closed loop in crystal containing the dislocation: it determines the magnitude and direction of slip in the crystal, and is the most characteristic feature of the dislocation. The Burgers vector of the dislocation is always the same and is independent of the circuit followed around the dislocation; we denote it $\boldsymbol{b}$ and its scalar magnitude b.

In the 3-dimensional crystal, the extra half-row of atoms shown in Fig. 1.7 becomes an extra half-plane of atoms (Fig. 1.8a), which extends indefinitely into the crystal in a direction perpendicular to the slip direction. For this particular configuration, the Burgers vector is perpendicular to the dislocation line, which is referred to as an *edge dislocation*. Dislocation lines with Burgers vector parallel to the line of the dislocation also exist. These are known as *screw dislocation*. The structure of the screw dislocation can be understood by reference to Fig. 1.8b, in which part of the crystal above the slip plane has been displaced by one interatomic spacing relative to the crystal underneath the slip plane. The description as a screw dislocation becomes clear when the nature of the atomic distortions around the dislocation are examined by viewing along the dislocation line. A traverse from atom A to B, C, D, E in that order penetrates the crystal approximately one lattice spacing inwards from A, whilst in a perfect crystal such a circuit would close at D. The form of the entire circuit through the crystal, shown in Fig. 1.8c,

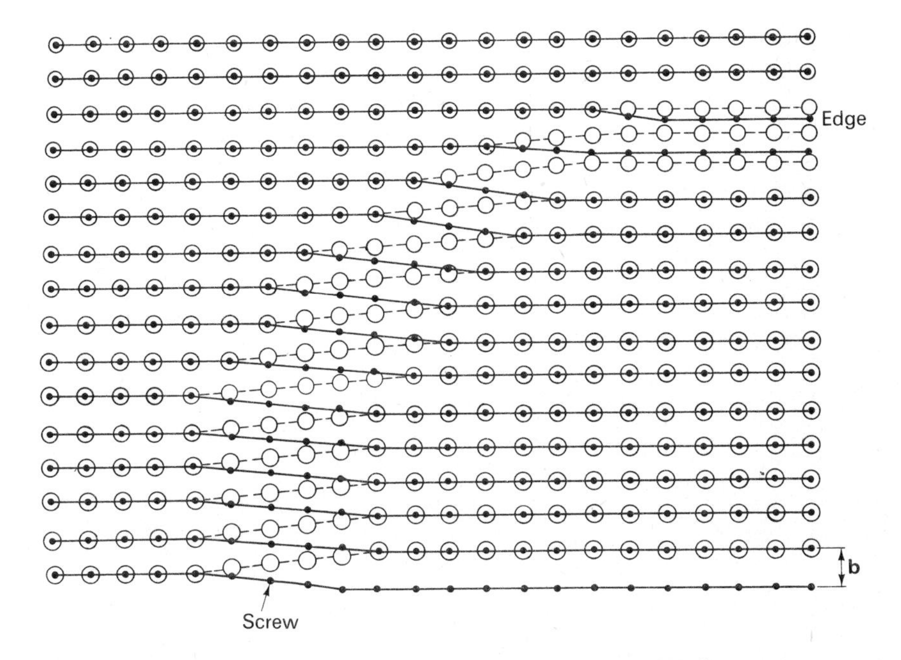

Fig. 1.9 The structure of a mixed dislocation in a cubic crystal. (Circles represent atoms in the plane above the slip plane, dots represent atoms in the plane below the slip plane.) The Burgers vector of the general dislocation may be resolved into components which are the pure edge and pure screw dislocations in the special orientations parallel to the cube edges. (After READ, W. T. 1953. *Dislocations in Crystals*. McGraw-Hill, New York.)

clearly resembles the form of a right handed screw. Whenever such a dislocation intersects a crystal surface, a step is produced at the crystal surface as shown in Fig. 1.8b. The important difference between the motion of the screw and edge dislocations under stress is clearly demonstrated in Fig. 1.8d. The screw dislocation moves perpendicular to both its own length and to the direction of slip, whereas an edge dislocation moves perpendicular to its length but parallel to the direction of slip. Since the screw dislocation has no extra half-sheet of atoms, there is no non-conservative motion and the dislocation is free to move on any cylindrical surface having the slip direction for its axis. In this sense the motion of the screw dislocation is less restricted than the edge dislocation.

It must be realized that the edge and screw dislocations described above are ideal dislocations, and that dislocations in real solids are of mixed character, part edge and part screw. The arrangement of atoms near a segment of a dislocation loop is shown in Fig. 1.9: in the special orientations parallel to the sides of the crystal there are pure edge and pure screw components. In general however, slip occurs by movement of all parts of the dislocation loop; edge, screw and mixed components.

Dislocation energies

The simple dislocation structures described above show that each dislocation has an associated strain field. Despite the severe local distortion around the dislocation, there is still matching of the lattice planes on either side of the dislocation. At the centre of the dislocation the strains are too large to be regarded as accurately Hookean displacements. The strains do decrease with increasing distance from the dislocation, however, such that displacements of atoms a few atomic spacings away from the dislocation may be treated using linear elasticity theory. In the region outside the dislocation core the strain energy may be computed with reasonable accuracy. Consider the presence of a screw dislocation in an elastically isotropic medium. It is evident that the screw dislocation has a different strain pattern than that of the edge dislocation, since it takes the form of a right or left handed screw of pitch equal to the Burgers vector. The elastic distortion, i.e., the distortion at some suitable distance from the core of the dislocation, can be represented by making a radial slit parallel to the axis of a hollow cylinder of material, which is itself parallel to the dislocation line. The free surfaces are then displaced rigidly with respect to one another in the manner of Fig. 1.10, by an amount b in the Z-direction. Figure 1.10 shows this displacement parallel to the dislocation caused by a simple shearing in a thin cylinder of radius r and length l. Evidently this shear strain is symmetrical about and directed along the dislocation line. Provided that the distortion is distributed uniformly over the circumference of the cylinder, we can define the shear strain ϵ and shear stress σ as

$$\epsilon = b/2\pi r \quad \text{and} \quad \sigma = \mu b/2\pi r$$

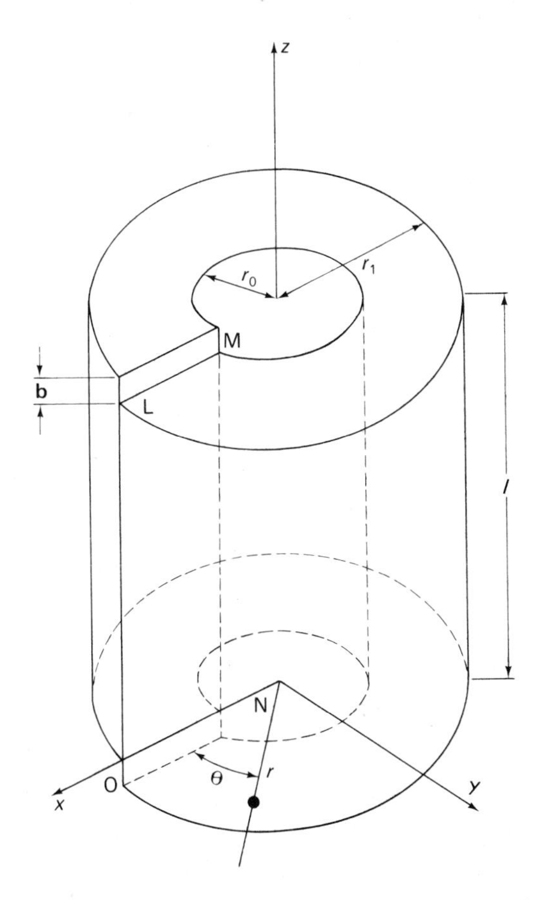

Fig. 1.10 A thin shell of elastically distorted material around a simulated screw dislocation. (After HULL, D. 1968. *Introduction to Dislocations*. Pergamon, Oxford.)

The strain energy density dU stored in this thin shell which has thickness dr and volume dV is given by,

$$dU = \tfrac{1}{2}\epsilon\sigma \, dV$$

where $dV = 2\pi lr \, dr$

$$U = \int_{r_0}^{r_1} \frac{\mu b^2 l}{4\pi} \frac{dr}{r}$$

Thus

$$U = \frac{\mu b^2 l}{4\pi} \ln \left(\frac{r_1}{r_0}\right) \tag{1.7}$$

where r_1 is the radius of the specimen. The magnitude of r_0 depends upon the displacements at the dislocation core, since r_0 represents the radius around the dislocation within which elasticity theory is invalid. The integration is in principle taken out to the crystal boundaries, and consequently equation 1.7 represents the total strain energy of the whole strain field of the dislocation, i.e., over the entire crystal volume. However, to estimate the magnitude of U, consider a crystal in which $r_o \simeq 10^{-7}$ cm, $r_1 = l = 1$ cm and $b = 3 \times 10^{-8}$ cm with $\mu = 5 \times 10^{11}$ dyne cm^{-2}. The resulting strain energy is about 7 eV per atom length of the dislocation line. It should be emphasized that more than one half of this energy is stored in the volume outside $r = 10^{-4}$ cm. This long-range strain field around the dislocation is of the utmost importance in influencing the interactions between dislocations and other defects. A similar computation of the strain energy of the edge dislocation gives the result,[2]

$$\boxed{U = \frac{\mu b^2 l}{4\pi(1 - \nu)} \ln \left(\frac{r_1}{r_0}\right)} \tag{1.8}$$

where ν is Poisson's ratio. Very roughly the energy stored in the core of the dislocation amounts to only a 10 to 20% correction to the energy stored outside the dislocation core[2]. There are two important further consequences of the high strain energy associated with the dislocation. Firstly, because of its high energy, the dislocation line will try to attain a lower energy by shortening its length. Thus the dislocation may be considered to possess a line tension. Secondly, despite the increased configurational entropy consequent upon their being present in crystals, dislocation lines increase the free energy of the crystal by an amount almost equivalent to the strain energy. Since the strain energy is very large, dislocation lines cannot exist in thermal equilibrium in solids.

For both edge and screw dislocations equations 1.7 and 1.8 show the dislocation line energy to be proportional to the square of magnitude of the Burgers vector $|\boldsymbol{b}|$ of the dislocation. If a dislocation can, therefore, divide into two (or more) partial dislocations, with Burgers vectors $\boldsymbol{b}_1$ and $\boldsymbol{b}_2$, such that $|\boldsymbol{b}_1|^2 + |\boldsymbol{b}_2|^2 < |\boldsymbol{b}|^2$ then the dislocation line can lower its energy. Such a tendency to dissociate into two or more dislocations having shorter Burgers vectors occurs in numerous structures. In the face-centred cubic structure the shortest vector in the slip direction is the $a/2$ [110] vector. This corresponds to the Burgers vector of the unit dislocation in the face-centred cubic lattice. One such vector $a/2\ [10\bar{1}]$ is shown in Fig. 1.11a representing slip in the (111) lattice plane. However, an atom resting in a B position on top of the A plane may find an easier movement to a neighbouring B position by following a zig-zag path via the C position. The end result, a displacement measured by the vector $\boldsymbol{b}$, is accomplished by the unit dislocation dissociating into two half dislocations with Burgers vectors $\boldsymbol{b}_1$ and $\boldsymbol{b}_2$ according to the reaction,

$$\frac{a}{2}\,[10\bar{1}] \rightarrow \frac{a}{6}\,[2\bar{1}\bar{1}] + \frac{a}{6}\,[11\bar{2}]$$

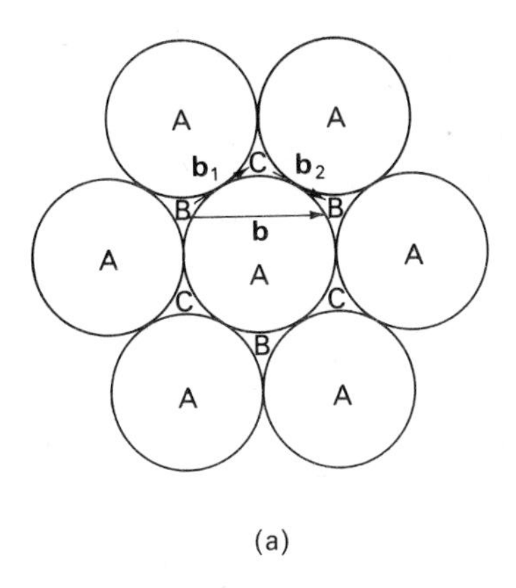

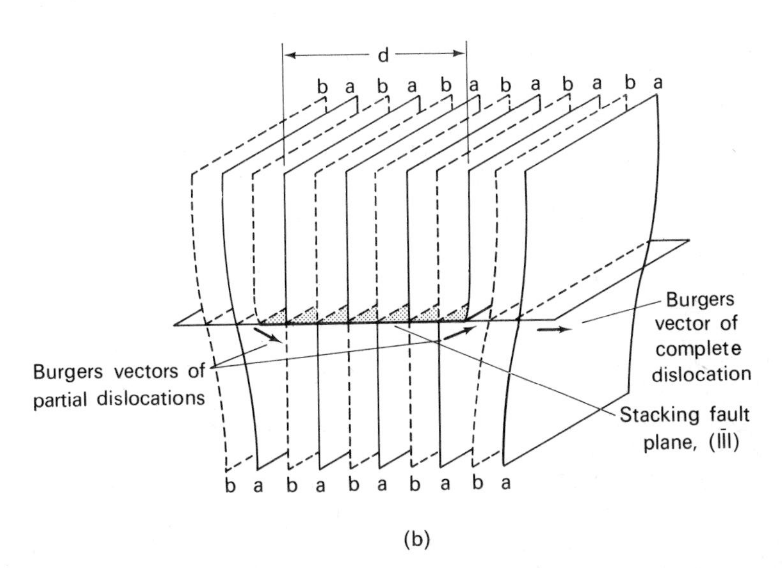

Fig. 1.11 (a) The nature of slip on the (111) slip plane of a face-centred cubic crystal, (b) two partial dislocations separated by a stacking fault on the (111) plane. The extended dislocation in a face-centred cubic crystal is formed by the dissociation of a unit edge dislocation $\mathbf{a}[10\bar{1}]/2$ into two partial dislocations with Burgers vector $\mathbf{a}[2\bar{1}\bar{1}]/6$ and $\mathbf{a}[11\bar{2}]/6$. (After SEEGER, A. 1957. *Dislocations and Mechanical Properties of Crystals.* Wiley New York.)

Such a dissociation is energetically favourable since the sum of strain energies (or $b_1^2 + b_2^2$) for the pair of half dislocations is $a^2/6 + a^2/6$, which is less than the energy of the single unit dislocation $a(^2/2)$. The strain field of these two partial dislocations will cause them to repel one another and separate on the slip plane, producing a sheet of *stacking fault* between them (Fig. 1.11b). It is the energy of this sheet of stacking fault which prevents them separating too far: the equilibrium separation of the partial dislocations can be estimated by equating the repulsive energy between them with the stacking fault energy. Since the Burgers vectors $\boldsymbol{b}_1$ and $\boldsymbol{b}_2$ are oriented at 60° to one another, they repel each other with a force that is approximately $(\mu \boldsymbol{b}_1 \cdot \boldsymbol{b}_2)/2\pi d$, where d is the separation between them. This repulsive force is balanced by the fault energy, γ in J cm^{-2} which pulls the dislocations together with a force of γ dynes. Hence the equilibrium separation is

$$\boxed{\mathrm{d} = \frac{\mu a^2}{24\pi\gamma}} \tag{1.9}$$

Thus the width of the stacking fault increases inversely as the stacking fault energy, γ. It is well established experimentally that in materials with low stacking fault energy (e.g., Cu, Au, stainless steel), the equilibrium separation of the partials may be of the order of 5 to 10 lattice spacings. Such separations may be much greater in the presence of an applied stress, as shown by the extended dislocation B in the electron micrographs of stainless steel in Figures 1.5c and 1.5d. The stacking fault is clearly revealed as a number of striped parallel bands or fringes. (This appearance of the fault in the electron microscope results from the phase difference between the electrons scattered above and below the fault, which is manifest as a change in the intensity of the diffracted electron beam.) Stacking fault B traverses the whole of a grain in the foil, and presumably the front and rear partial dislocations are pinned at the grain boundaries. Figures 1.5c and 1.5d also show that the fault A widens as a result of stresses set up near the fault. Since generally the partial dislocations and their intervening stacking fault move as a unit on the (111) slip plane, the trailing partial dislocation in fault A must be pinned. In nickel or aluminium where γ is high, the dislocations are essentially unextended. However, no matter how wide the separation of the partials, the two dislocations must move together across the slip plane since they are bound together by the stacking fault. In fact the basis for the remarkable plasticity of some metals is that the interatomic forces of crystals offer practically no resistance to the motion of a wide dislocation.

Interaction between dislocations: tilt boundaries and grain boundaries

We have shown that both edge and screw dislocations have an associated long-range elastic strain field. Thus dislocations will interact with one another

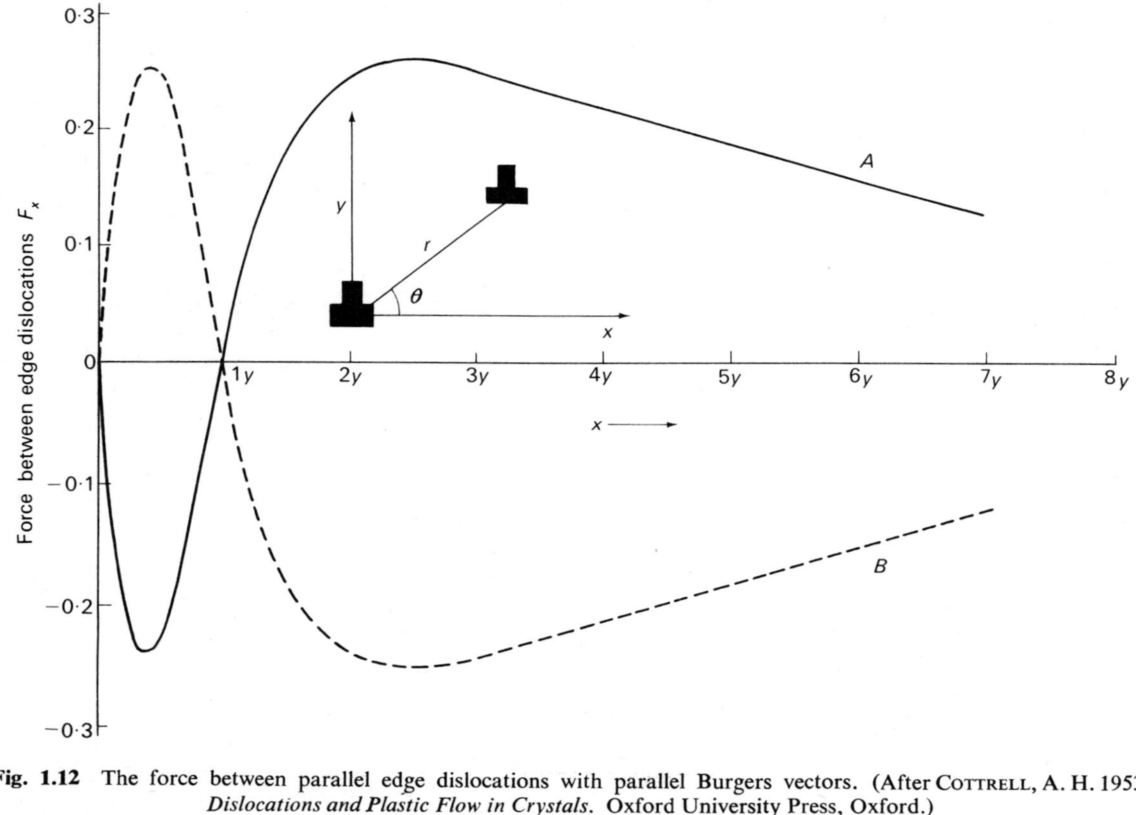

Fig. 1.12 The force between parallel edge dislocations with parallel Burgers vectors. (After COTTRELL, A. H. 1953. *Dislocations and Plastic Flow in Crystals.* Oxford University Press, Oxford.)

even when widely separated, so as to assume positions of minimum energy and reduce the total strain energy of the lattice. For example, a positive and a negative edge dislocation on the same slip plane will be attracted to one another in order that their respective strain fields cancel. However, if the two dislocations have parallel Burgers vectors, they will repel each other with a force which decreases inversely as their separation. Making use of Equ. 1.8, we see that the force on unit length of dislocation along the line joining the two dislocation cores is just

$$F = \frac{\mu b^2}{2\pi(1 - \nu)r}$$

Of more importance, however, is the force between two edge dislocations with parallel Burgers vectors moving on parallel slip planes. This may be evaluated using the method of surface integration;[2] in our present discussion we merely state the principal results. The inset to Fig. 1.12 shows two edge dislocation lines along the Z-axis with their Burgers vectors along the X-axis. The two dislocations are joined by a line of length r which makes an angle θ with the X-axis. The radial and tangential components of the force are given by,[2]

$$F_r = \frac{\mu b^2}{2\pi(1 - \nu)r} \quad \text{and} \quad F_\theta = \frac{\mu b^2 \sin 2\theta}{2\pi(1 - \nu)r}$$

Since edge dislocations move only in the slip plane we need examine only the force acting in the X-direction. This is

$$F_x = F_r \cos\theta - F_\theta \sin\theta$$

and substitution for F_r, F_θ, and the trigonometric functions gives

$$\boxed{F_x = \frac{\mu b^2 x(x^2 - y^2)}{2\pi(1 - \nu)r^4}} \tag{1.10}$$

The important consequences of this equation are shown in Fig. 1.12, where we plot F_x as a function of x: for convenience y, the distance between the slip planes, is a scaling distance for x, equal to the unit of length along the x-axis. F_x is zero at both $x = 0$ and at $x = y$, but the equilibrium is stable at $x = 0$ (i.e., when one dislocation lies vertically above the other). For $x > y$ the two dislocations repel one another, unless they are of opposite sign. These conclusions are in general true when a large number of edge dislocations of the same sign is involved. Dislocations in such an array will tend to arrange themselves in a plane normal to the slip plane. In these regions of high dislocation density the lattice will be bent through a small angle. For this reason they are referred to as *small-angle* or *tilt boundaries*.

A simple tilt boundary consisting of an array of edge dislocations in a cubic crystal is shown in Fig. 1.13a. The dislocation lines are arranged parallel to the [010] axis and are equally spaced at intervals D along this axis. The misorientation may be described by a small rotation θ about the [001] axis common to the material on either side of the boundary. Each dislocation in the boundary will experience the strain fields of the dislocations above and below it, in addition to its own strain field. However, since

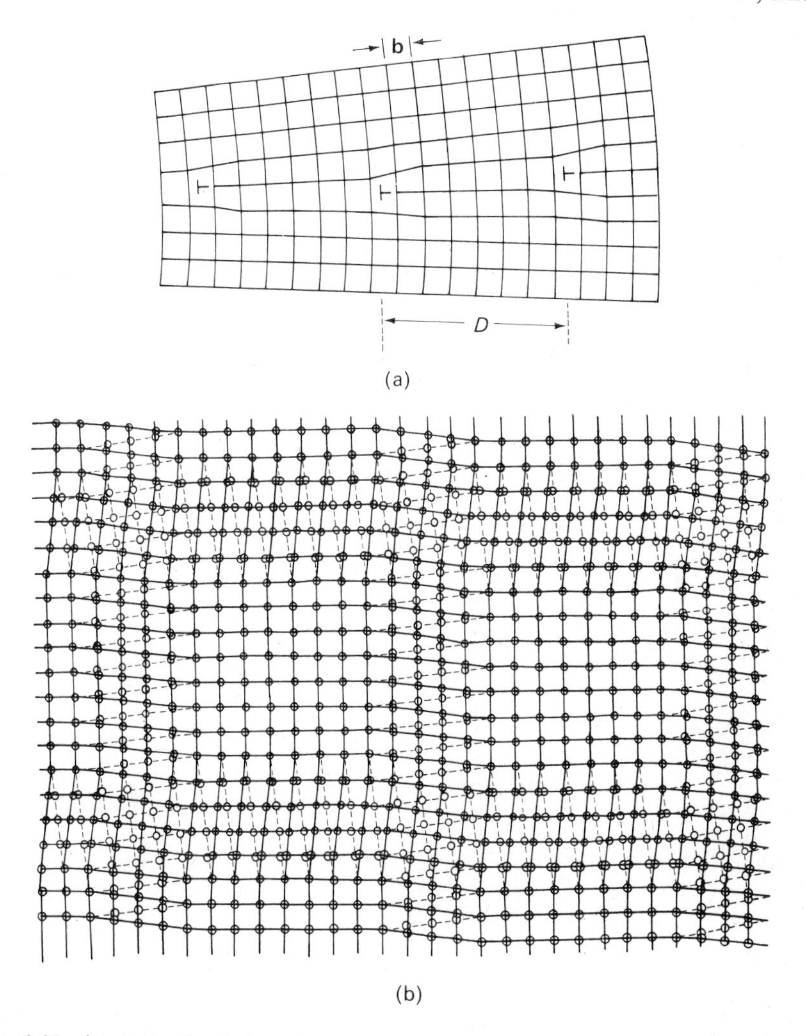

Fig. 1.13 (a) A simple tilt boundary (after VOGEL et al. 1953. *Phys. Rev.* **90,** 489) and (b) a pure twist boundary. (After READ, W. T. 1953. *Dislocations in Crystals.* McGraw-Hill, New York.)

the strain fields above and below a dislocation line are equal and opposite, the strain fields from other dislocations almost cancel. Thus the strain energy of each dislocation in the boundary results primarily from its own strain field. The radial force on the dislocation is given above as

$$F_r = \frac{\mathrm{d}E}{\mathrm{d}r} = \frac{\mu b^2}{2\pi(1-\nu)r}$$

Now any incremental change $\mathrm{d}r$ in r will correspond to an increase $\mathrm{d}D$ in D and a corresponding decrease $-\mathrm{d}\theta$ in the rotation angle θ. Consequently since $-\mathrm{d}\theta/\theta = \mathrm{d}D/D = \mathrm{d}r/r$, we find that the energy per dislocation is

$$E = -\int \frac{\mu b^2}{2\pi(1-\nu)} \frac{\mathrm{d}\theta}{\theta}$$

$$= -\frac{\mu b^2}{2\pi(1-\nu)}(A + \ln\theta)$$

where A is a constant.

For unit length of boundary there are $1/D$ dislocations present in the boundary, and at small angles $1/D = \theta/b$. Thus the energy of unit area of the boundary is given by

$$\gamma = \frac{\mu b}{4\pi(1-\nu)}\theta(A - \ln\theta) \tag{1.11}$$

Thus the tilt boundary energy is zero when θ is zero; as θ increases γ rises to a maximum after which it decreases. Surprisingly, this equation seems to account for experimental observations in tin, lead and silicon-iron even at large angles ($\theta \simeq 30°$).[4] The direct verification of the dislocation model of tilt boundaries has been provided by optical measurements, in which the number of dislocations present in a boundary correlated well with the angle of tilt measured using X-rays. It is further supported by the fact that pure tilt boundaries move normal to themselves upon application of an applied stress.[4,5] The process of dislocation alignment occurs when severely bent crystals are annealed to relieve the internal stresses: it is usually referred to as *polygonization*.

Because the equation for γ involves a term $-\ln\theta$, two tilt boundaries may decrease their total energy by forming one boundary of greater misorientation. It is clear that if such a process were to occur on a wide scale *grain boundaries* would result. Lomer and Nye[6] have studied the change in structure of the tilt boundaries with increasing angle. At high angles there was apparently a large vacancy concentration in the boundaries, and such boundaries are better described by Mott's[7] model of a grain boundary, which involves a region of the lattice in which a good or bad fit exists between the two parts of a crystal. Essentially, the symmetrical tilt boundaries of both small and large angle are extremely mobile and may assist in long-range diffusion processes in solids. Unsymmetrical boundaries composed of more than one family of slip planes are comparatively immobile. Finally, we comment that low angle *twist boundaries* are produced when at least two sets of parallel screw dislocations interact with one another. Such a boundary is illustrated schematically in Fig. 1.13b. For arrays of parallel screw dislocations to be stable there must be more than one set. One alone is ordinarily unstable as it produces a high shear strain extending a long way from the boundary. The second set of parallel screw dislocations tends to cancel this far-reaching shear strain, so forming a stable boundary.

Appendix: Thompson's Reference Tetrahedron

A simple construction due to Thompson (1953, *Proc. Phys. Soc.*, B, **66**, 481) greatly assists an understanding of interactions between dislocations.

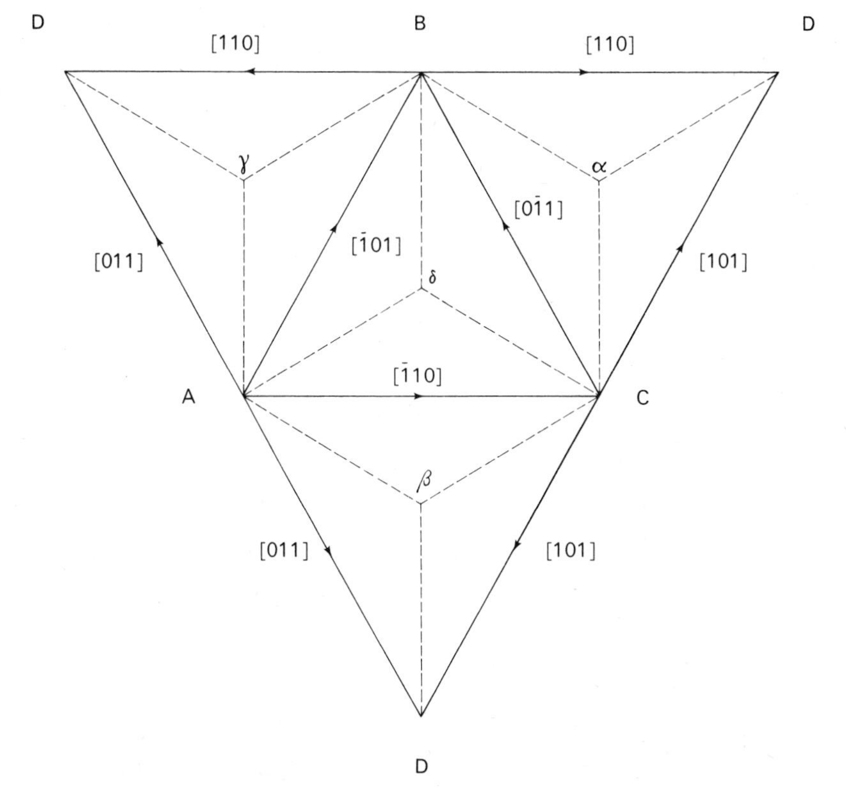

Fig. A1 Thompson's Reference tetrahedron. The equilateral triangles ABC, ADC, ADB and BCD represent the crystallographic planes (111), (11$\bar{1}$), (1$\bar{1}$1) and ($\bar{1}$11) in the face-centred cubic lattice.

The tetrahedron ABCD in Fig. A1 is made up of four $\{111\}$ planes; for convenience this tetrahedron is opened out, the vertex D having originally been above (or below) the plane of the paper. Consequently the edges AB, BC, CD, DA etc. correspond to the $\langle 110\rangle$ directions in the appropriate planes, and may be used to represent the undissociated slip vectors $(a/2)\ \langle 110\rangle$. There are twelve such vectors representable by the six edges of the tetrahedron each taken in two senses. Lines joining the vertices of the tetrahedron to the mid-points α, β, γ and δ of the faces lie parallel to the $\langle 112\rangle$ directions in the crystal and correspond to the $(a/6)\ \langle 112\rangle$ Burgers vectors of the Shockley partial dislocations discussed in §1.3.2. Thus we use a simple notation AB, BC – – – –, Aβ, Bα, etc., to describe the Burgers vectors of all dislocations in the face-centred cubic lattice. The dissociation reaction

$$\frac{a}{2}\,[\bar{1}01] \rightarrow \frac{a}{6}\,[\bar{2}11] + \frac{a}{6}\,[\bar{1}\bar{1}2]$$

can now be written in the convenient shorthand

$$\mathrm{AB} \rightarrow \mathrm{A}\,\delta + \delta\,\mathrm{B}$$

Another important partial dislocation, the Frank sessile dislocation (see §6.2.2) has its Burgers vector normal to plane containing it. For a dislocation in the (111) plane, the Burgers vector is $(a/3)\ \langle 111\rangle$ or αA.

Dislocation reactions in hexagonal closepacked lattices may be pictured by a representation similar to the Thompson tetrahedron (see BERGHEZAN, FOURDEUX and AMELINCKX, 1961. *Acta Met.* **9**, 464).

GENERAL REFERENCES

COTTRELL, A. H. 1953. *Dislocations and Plastic Flow in Crystals.* Oxford University Press, Oxford.

MOTT, N. F. and GURNEY, R. W. 1940. *Electronic Processes in Ionic Crystals.* Oxford University Press, Oxford.

READ, W. T. 1953. *Dislocations in Crystals.* McGraw-Hill, New York.

SCHULMAN, J. H., and COMPTON, W. D. 1962. *Colour Centres in Solids.* Pergamon Press, Oxford.

VAN BEUREN, H. G. 1960. *Imperfections in Crystals.* North-Holland Publishing Co., Amsterdam.

2

Some Experimental Techniques

The physical size of a defect usually decides the experimental techniques needed to discern its structure and properties. Point defects are so small as to prohibit their observation directly by microscopic techniques. Instead indirect methods are used which compare some convenient property of a crystal containing defects with the same property in *defect-free* crystals. In some cases the techniques of electron spin resonance and electron nuclear double resonance have given a very detailed picture of the point defect and its environment. Dislocations and stacking faults are several orders of magnitude larger than are point defects, and their properties may be investigated in the optical and electron microscope. The following discussion of experimental techniques used in the study of defects is by no means comprehensive. It is intended merely to emphasize the physical principles involved in studying the defect solid state.

2.1 The production of defects in solids

In order to study the properties of defects in crystals it is necessary that they be present in a sufficiently high concentration. A number of convenient techniques are used to increase the concentration of defects above the thermal equilibrium concentration. The principal methods include quenching from high temperature, irradiation with energetic particles, alloying and plastic deformation.

Quenching from high temperature

It was shown in Chapter 1 that a real crystal contains intrinsic point defects in thermal equilibrium in concentrations determined by the Boltzmann factor $\exp(-E/kT)$. Consequently the point defect concentration increases rapidly with increasing temperature and in some solids approaches 10^{-3} near the melting point. This is the case for aluminium where the vacancy formation energy is only $\sim 0{\cdot}7$ eV. Neglecting the temperature independent term related to the vibrational entropy changes, we find that at 900 K,

$$\frac{n}{N} = \exp - \left(\frac{0{\cdot}7}{8{\cdot}62 \times 10^{-5}} \cdot \frac{1}{900}\right) = \exp(-9) \simeq 10^{-4}$$

since 1 K $= 8{\cdot}62 \times 10^{-5}$ eV. If the crystal is cooled slowly from a high temperature, the point defects will migrate to lattice positions at which they are annihilated in order that thermal equilibrium be maintained. Rapid cooling or *quenching* from high temperature precludes this process and consequently a supersaturation of vacancies exists at the lower temperature. Since the energy required to form vacancies is generally lower than that for interstitial formation, quenching is a more suitable means of producing vacancies than interstitials. Vacancies produced in this way are important ingredients in the kinetics of many solid state processes, especially those such as precipitation hardening where atomic diffusion is involved.

In general, quenching rates are insufficiently rapid to retain completely at low temperature the excess defect concentration formed at higher temperatures. It is most nearly possible to retain the defect-supersaturation in metals, where the high thermal conductivity ensures rapid quenching. There is, however, usually a tendency for dislocation loops to be spontaneously nucleated in metals during quenching. The driving force for the process is the *excess free energy* of the crystal, consequent upon the presence of a defect supersaturation. This is readily appreciated when it is recalled that in calculating the equilibrium defect concentration, $(\partial F/\partial n)_T$ was equated to zero (p. 10). On quenching the crystal through the temperature range $T_2 - T_1$, we retain at T_1 more than the equilibrium concentration of vacancies so that $(\partial F/\partial n) \neq 0$. Actually for Schottky defects at T_1 K

$$\frac{\partial F}{\partial n} = E_S + kT_1 \ln \left(\frac{n_2}{N + n_2}\right)$$

if the consequences of vibrational entropy changes are neglected. However $E_S = -kT_1 \ln \left(\frac{n_1}{N + n_1}\right)$, so that

$$\frac{\partial F}{\partial n} = kT_1 \ln \left(\frac{n_2}{n_1}\right)$$

Now ∂n defects introduce a change of volume ∂V into the crystal, and if L^3 represents the volume per vacancy we find that

$$\frac{\partial F}{\partial V} = \frac{kT_1}{L^3} \ln \frac{n_2}{n_1}$$

Using Equ. 1.1 to find $\ln\left(\frac{n_2}{n_1}\right)$, we see that

$$\frac{\partial F}{\partial V} = \frac{E_S}{L^3}\left[1 - \frac{T_1}{T_2}\right] \tag{2.1}$$

Since stress = energy/unit volume it is evident that the supersaturation of vacancies exerts an effective internal stress. For aluminium where E_S = 0·7 eV, this gives an effective stress of 300 kg mm^{-2}, far above the theoretical yield stress of aluminium. This effective stress can be relieved by migration to grain boundaries and dislocations; of somewhat greater importance, however, is the spontaneous nucleation of dislocation loops discussed in §1.1 and §6.2.

Defect production by irradiation

Defect production by energetic particle irradiation depends not only upon the nature and energy of the incident particle, but also upon the bonding and crystal structure of the irradiated solid. The incident particle shares its energy with an atom in the solid, which may be ejected from its normal lattice site, leaving behind a crystal vacancy. Thus the simultaneous products of irradiation are interstitial atoms and vacant lattice sites. The energy required to displace atoms in this way is related to the strength with which an atom is bound to its lattice site. The displacement energy, E_D, represents the sum of the energy required to break the bonds* with neighbouring atoms, the energy required to overcome the opposition of the lattice to the movement of the displaced atom, and the energy regained in establishing bonds at whatever lattice site the atom comes to rest. Intuitive reasoning suggests that the struck atom must acquire 25 eV of energy in a collision before it can be displaced from a normal lattice site into an interstitial position. If the energy transferred to the struck atom is less than this the atom is not displaced from its lattice position, but instead dissipates the energy acquired in the collision through vibrational motion. Atoms which are displaced are referred to as *primary knock-ons.* This simple picture of the displacement process implies that the displacement energy has a discrete value in crystalline solids. If such were the case the probability of a collision resulting in atomic displacement would be zero for all energies below E_D, and unity for all energies greater than E_D. In this case the probability displacement curve is the step function shown in Fig. 2.1. Thermal vibrations broaden this step function quite appreciably, although more significantly consideration must be given to the effects that the ordered crystal lattice has upon the displacement

* This terminology should not be interpreted too literally; it is however helpful sometimes to picture the binding energy of an atom in a lattice as shared between the conceptual "bonds" of the atom to its neighbours.

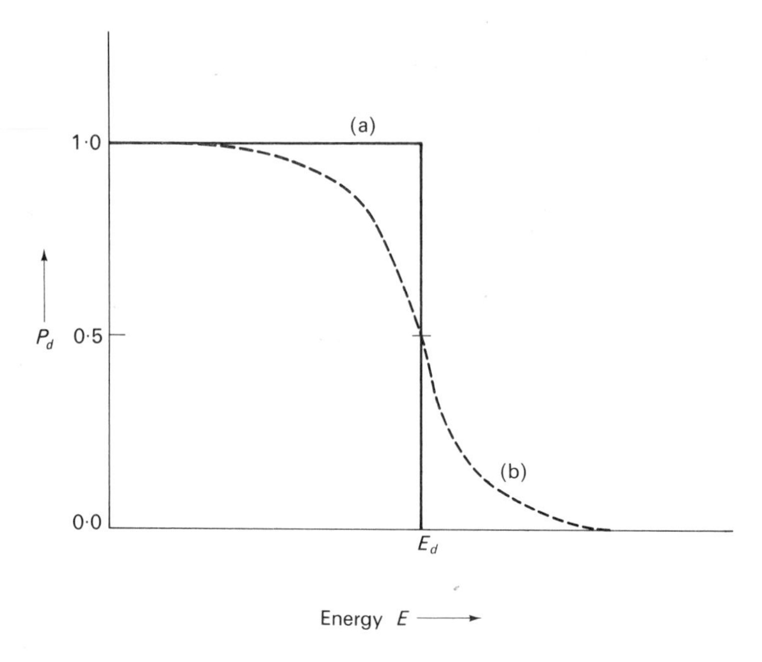

Fig. 2.1 The displacement probability, p_d, as a function of the kinetic energy transferred to a lattice atom. In (a) a sharp displacement threshold is assumed. (b) takes account of such effects as crystal anisotropy and temperature, both of which modify the step function.

process. It is quite unrealistic to expect that an atom ejected towards a position in the lattice between two neighbouring atoms will experience a greater repulsion than if it were involved in a head-on collision with a neighbouring atom. Thus the displacement energy is a function of both temperature and crystal symmetry. Taking account of these effects will change the displacement probability curve, as shown schematically in Fig. 2.1. In this idealized displacement probability curve, E_D is interpreted as the energy transferred to an atom at which the probability of displacement in some arbitrary direction and at a given temperature is $\frac{1}{2}$. Typical values of E_D are 20 eV for Cu, 28 eV for Ag, 40 eV for Au, 35 eV for Ni and 31 eV for Ge. Subsequent collisions between the incident particle and other atoms in the crystal may also cause displacements. Such further primary events occur with high probability as long as the radiation is sufficiently energetic that energy greater than E_D can be transferred to the struck atom in the collision. If energy less than E_D is gained by the atom in a collision, it is dissipated in thermal vibrations.

Since the primary knock-ons are frequently extremely energetic they cause further events as they move through the lattice. The most important higher order processes include electron excitation and ionization together with the production of secondary, tertiary, and subsequent knock-ons. Energy losses by the primary knock-ons result in electron excitation and ionization only when the knock-on energy exceeds the ionization threshold or limit, E_C. Below this limiting energy all energy losses are due to collisions with other lattice atoms. The knock-ons will continue to cause displacements only whilst their energy is sufficient for them to transfer energy greater than E_D to the atoms with which they collide. Thus the process of defect production during irradiation involves the incident radiation in sharing its energy with the lattice atoms, causing some of them to be displaced from normal lattice sites. The energy of the displaced atoms is similarly degraded by interactions with other atoms until the energy transferred in a collision is less than E_D. The displaced atoms then come to rest in a interstitial sites, dissipating their remaining energy by exciting lattice vibrations.

The exact nature and magnitude of the displacement process is not yet fully understood. It is generally believed that the *displacement spike* of a single 2 MeV neutron may involve a region of the crystal containing as many as 10^5 atom sites, at the centre of which are lattice vacancies with the interstitials being closer to the periphery of the *spike*. Owing to the thermal motion of the atoms many vacancy-interstitial pairs spontaneously recombine, so reducing the amount of damage in the region of the displacement spike. The local concentration of defects is typically of the order of $10^{-4} - 10^{-3}$.

We noted above that atoms coming to rest following displacement from their normal lattice sites may dissipate energy in thermal excitations of the lattice. Energy disposed of in this way may amount to 25 eV per displaced atom. Although the energy is deposited in only a small region of the crystal and persists for only a short time ($10^{-11} - 10^{-12}$ sec), it may profoundly alter the distribution of radiation damage in the crystal. Such regions of local heating are referred to as *thermal spikes* to distinguish them from displacement spikes. Since a metal like copper requires only about 0·25 eV per atom to raise its temperature to the melting point, these short pulses of energy could cause local melting for a region of the lattice containing 100 atoms. In view of their short duration, it is improbable that equilibrium be attained within the affected volume or that actual melting takes place. Nevertheless thermal spikes do assist in such phenomena as radiation annealing, especially in heavily irradiated solids where displacement cascades overlap.

Numerous irradiation sources produce defects in crystalline solids. Anion displacements in the alkali halides are frequently low energy events and ionizing radiations may be used. The term *ionizing radiation* describes radiation sources which generate free electrons or holes when interacting with matter. Thus ultra-violet light with energy $\sim$10 eV, 10–100 keV x-rays, 1·25 MeV γ-rays and high energy (100 keV – 10 MeV) electrons are all examples of ionizing radiation. High energy protons (2-20 MeV) are a further example of ionizing radiation: in view of their large mass, energetic protons also create defects by direct momentum transfer in collisions. With the exception of γ-rays, these sources of irradiation are not especially penetrative, but for high doses large sources are necessary and extensive radiological protection facilities are necessary. In the easily damaged alkali halides, ultra-violet radiation and especially soft x-rays are the preferred sources of irradiation. In metals and the more covalent solids, more energetic irradiation sources are necessary to produce detectable levels of damage.

The behaviour of solids under reactor irradiation is of considerable importance to technologists. The interest is in the design of nuclear reactors, in which both the fissile and non-fissile components of the reactor core must maintain structural integrity, dimensional stability and mechanical strength during reactor operation. In reactor cores the ingredients of the fission flux include neutrons of varying energies, fission fragments, x-rays and electrons. The overwhelming majority of displacement events in reactor irradiated solids are initiated by primary events involving neutrons colliding with atomic nuclei. A spectrum of energies is available for neutrons in the reactor pile. *Thermal neutrons* are in thermal equilibrium with their environment: at 18°C their mean energy ($\sim$1/40 eV) is insufficient to cause atomic displacements. Neutrons with energies greater than 0·2 keV are referred to as fast neutrons: energies greater than 2 MeV are produced by fissile reactions in the reactor. Since they are uncharged, neutrons are unaffected by the electrical fields around atomic nuclei and consequently they can travel large distances in a solid. The resultant damage is not, therefore, restricted to the surface region but is distributed uniformly throughout the crystal.

The theories of atomic displacement in solids are well documented in the literature. We do not reproduce the arguments here. However, we do note that estimates of the intensity of radiation damage are usually three to five times in excess of that determined experimentally. This is because of a number of unjustified simplifications made in the theoretical models. In addition, the models usually neglect effects due to the discrete lattice structure, the probability of vacancy-interstitial recombination, and radiation annealing, all of which tend to reduce the amount of radiation damage. These calculations are described fully in the references[8–10].

Defect production by alloying

An excess of lattice vacancies may be formed in some metallic solid solutions, in which the concentration of unoccupied lattice sites changes significantly with alloy composition. The outstanding examples are the face-centred cubic solid solutions in the aluminium-zinc, aluminium-magnesium, and gold-nickel systems. Unsuccessful attempts have been made to correlate the vacancy concentration dependence on alloy composition with the Brillouin zone structure of the alloys. Similar effects are observed in the intermetallic compounds Cu_9Al_4, Cu_9Ga_4, and the transition metal aluminides, NiAl, CoAl and FeAl. For these compounds, however, it is found that the extent to which the atoms are omitted from the lattice corresponds to the maintenance of a constant number of electrons per unit cell. A survey of the available data suggests that this critical electron concentration is a limit beyond which the normal structure cannot exist. A modification of this type of *non-stoichiometry* is to be found in ionic solids, such as the alkali halides, the alkaline earth oxides and halides and numerous oxides with the rutile structure. More complex structures also show non-stoichiometry.

A convenient means of introducing excess lattice vacancies into the alkali halides is the technique of so-called additive colouration. This process amounts to heating the alkali halide crystal to a high temperature in the vapour of either alkali or halide constituent. The colouration is easiest for the alkali metal, although care must be taken to avoid impurity contamination. In addition it is necessary to cool the sample rapidly in order to prevent aggregation of the vacancies. The excess of alkali metal in the crystal results in the formation of halide ion vacancies at which electrons released from the metal atoms become trapped. The resulting complexes, anion vacancies each containing one electron, are called *F*-centres. These defect-containing crystals are necessarily electrically neutral. An excess of the halide constituent is more difficult to achieve, vapour pressures of greater than 50 atmospheres being necessary together with very high temperatures. Additive colouration techniques have also been used successfully with alkaline earth halides and oxides. Defect structures in some non-stoichiometric oxides (e.g., TiO_2, UO_2, PuO_2, NiO, FeO) lead to relatively wide ranges of composition being stable.

Defects introduced by plastic deformation

In general, defect lattice sites may be introduced at the intersections of dislocations. Experimental studies show that the concentration of vacancies generated in this way is related directly to the amount of plastic deformation.

2.2 Experimental methods in the study of defects

The experimental methods that are most generally applicable to the study of defects may be divided into three categories. Since lattice defects modify the properties of solids, important information about the defects may be inferred by monitoring the change in some convenient property as a function of a variable such as temperature. Included in this category are measurements of lattice parameter, density, thermal conductivity and electrical resistivity. A second category of experiment studies the response of the defect itself to the stimulus of some externally applied perturbation. The techniques of electron spin resonance and electron nuclear double resonance (ENDOR), optical absorption and fluorescence, and internal friction are examples of such experiments. The magnetic resonance and optical techniques often allow an unusually detailed description of the interaction of the defect with its environment, when the defect energy levels lie within the band gap of the host material. Such techniques are widely used in studying the defect structure of insulating solids and semiconductors. Direct observation of defects using microscopic techniques constitute the third category of interest. Microstructural investigations are especially useful since they are capable of direct resolution of the defect structure. In this respect magnetic resonance methods may be thought of as atomic microscopes, in view of the detail they reveal about the structure of defects.

2.3 Density and lattice parameter changes

The presence of defects in a crystalline lattice may change both the unit cell size and the physical size of the crystal. Consider first of all the Schottky defect. The presence of Schottky disorder in the lattice corresponds to atoms in the bulk crystal being removed to crystal surfaces. This will cause a decrease Δd in bulk density related to the number of Schottky defects, n_s, by

$$-\Delta d/d = n_s(V + v_s) \tag{2.2}$$

where V is the atomic volume and v_s is the relaxation around the vacancy. If the relaxation around the vacancy is significant then the lattice parameter of the crystal is expected to change. (This is analogous to the variation in lattice parameter of pure solids on alloying, the alloying addition here being the vacancy.) Obviously the sense of the density change associated with this relaxation depends upon whether the atoms neighbouring the defect move outwards or inwards. The change is usually small, for example, in magnesium oxide ENDOR measurements suggest that the nearest neighbour ions around an F-centre relax outwards by only 5 to 7% of the lattice constant. Consequently, it is usually assumed that no change occurs in the size of the unit cell, but only in the macroscopic dimensions of the crystal. Thus measurement of density and lattice parameter should reveal the inequality $\Delta d/d \gg 3\Delta a/a$.

When Frenkel defects exist both bulk density changes and unit cell changes are significant. The unit cell size changes because of the large dilation associated with the interstitial atom being squeezed into the relatively small spaces between atoms on normal lattice sites. In this case, and assuming that the Frenkel defects are randomly distributed and constitute centres of isotropic dilation in the crystal we may write,

$$\Delta a/a = -\tfrac{1}{3}\,\Delta d/d = \tfrac{1}{3}n_F(v + v_i) \tag{2.3}$$

In this relationship n_F is the number of Frenkel defects and v and v_i respectively represent the volume dilation around the vacancy and interstitial components of each Frenkel defect. When both Frenkel and Schottky defects are present some suitable summation of Equ. (2.2) and (2.3) is necessary. Clearly, comparison of the changes in macroscopic density and unit cell size can distinguish between the nature of the disorder present in the lattice.

Sufficiently accurate measurements of the unit cell and crystal density also permit the energy of formation of defects to be calculated. Both the crystal dimensions and the lattice parameter expand as a function of temperature the coefficient of expansion being same in both cases. Thus the difference ($\frac{1}{3}\,\Delta d/d - \Delta a/a$) assesses the contribution of defects to the dimensional changes of the crystal. Measurements of this difference as a function of temperature will be related to the activation energy for defect production through the Boltzmann relationship. Experimentally it is more convenient to measure simultaneously on the same sample the unit cell size and crystal dimensions. In this case the magnitude of ($\Delta L/L - \Delta a/a$) reveals the nature of the defects present in thermal equilibrium. An accuracy of at least 1 part in 10^5 is desirable in the measurements of both $\Delta L/L$ and $\Delta a/a$. Thus the measuring temperature must be constant to better than $\pm 0{\cdot}2°C$.

2.4 Electrical resistivity measurements

Lattice defects scatter both electrons and phonons and consequently contribute to both the electrical and thermal conductivity of solids. In electrical resistivity the electrons are scattered† by a perturbation in the regular lattice potential with a relaxation time τ_k, which involves a transition for the electron from a state designated by wavevector k into another state with wavevector k'. We can see the effect of this scattering process since the resistivity, ρ, is given by

$$\rho = m^*/ne^2\tau_k \tag{2.4}$$

where m^* and e are respectively the effective mass and charge of the electron[11]. Although n represents the density of electrons, the high conductivity of metals is due to the high velocity of electrons. Phonons scatter electrons since they set up perturbations in the regular lattice potential in various parts of the crystal. The lattice vibrations will, therefore, contribute an *intrinsic* resistivity ρ_i which we expect to increase as the temperature increases. Since each defect produces a perturbation in the lattice potential which is essentially independent of temperature, the *residual* resistivity per lattice defect is temperature independent. For low defect concentrations we may envisage the electrons as travelling through a perfect lattice, occasionally being scattered by the changed potential around each defect. Thus the scattering probability, and hence the residual resistivity, due to defects, vary directly as the defect concentration. Since the equilibrium defect concentration vary exponentially with temperature, we see that the total defect-induced resistivity ρ_0 is temperature dependent. Consequently the

† A detailed account of the scattering effects leading to electrical resistivity will be found in the companion book in this series on the *Electrical Properties of Solids*. See also.[11]

total resistivity, $\rho(T)$ of a solid may be represented by the sum of two temperature-dependent terms, viz:—

$$\rho(T) = \rho_0(T) + \rho_i(T) \tag{2.5}$$

The theory of electrical resistivity is extremely complicated, and the results are in only rough qualitative agreement with experiment. In fact the value of $\rho_i(T)$ at high temperature is usually extrapolated from low temperatures using, say, the Bloch-Gruneisen function (see, for example, the book referred to above). Since this function does not give a completely accurate representation of ρ_i, $\rho_0(T)$ obtained from high temperature must be somewhat uncertain. It is usual to try to overcome this by quenching from a succession of high temperatures, and making measurements at a single temperature. In this way $\rho_i(T)$ should be constant and $\rho_0(T)$ should be obtained with high accuracy, especially in metals, where the conductivity is easy to measure.

The foregoing discussion has been concerned only with the *electronic* conductivity, and is hardly relevant to conduction in insulating solids. In insulators conduction proceeds by the motion of lattice defects. However, since both the defect concentration and the motion are *thermally activated* processes which increase exponentially with temperature (see §3.1), we can use precisely the same methods for insulators as for metals in these cases. However in view of the small thermal conductivities of ionic solids, quenching methods are usually avoided.

It should be pointed out that by combining the three techniques discussed here, it may be possible to estimate the dilation around a defect. For example, a fractional volume decrease, given by $v_s/V = 0{\cdot}57$, has been measured for each Schottky vacancy in gold. Although these techniques have been used successfully in metals, semiconductors and insulators, there are many precautions which must be followed in experiments. Great care must be taken to avoid contamination during high temperature experiments. Furthermore, in view of the tendency for aggregate formation at high temperature, temperatures above about two thirds of the melting temperature should be avoided. Implicit in the quenching experiment is the assumption that the non-equilibrium situation at low temperatures entirely reflects the high temperature equilibrium situation. This is a highly suspect assumption since we showed earlier that it is reasonable to expect defect loops to be nucleated during quenching. There are a number of other techniques which have been used, including thermal conductivity and stored energy measurements. These are generally speaking less accurate and are not discussed further.

2.5 Electron spin resonance and electron nuclear double resonance

Electron spin resonance is limited to the study of defects having unpaired electrons, i.e., the total spin $S \neq 0$. Consider a small concentration of defects with one unpaired electron, distributed randomly in a diamagnetic environment and sufficiently well separated so that no appreciable magnetic interaction takes place between defects. In the absence of a magnetic field the energy levels of the electrons are two-fold spin degenerate with energy E_0. An applied magnetic field removes the spin degeneracy and the new energy levels are,

$$E_1 = E_0 \pm \tfrac{1}{2} g\beta H_0$$

g being the spectroscopic splitting factor, β the Bohr magneton and H_0 the magnetic field. Electron spin resonance is concerned with inducing magnetic dipole transitions between these spin levels. When such transitions do occur there is an absorption of energy, as depicted in Fig. 2.2a. Transitions are induced by applying an alternating magnetic field H_1 of frequency ν. An oscillating field in the direction of the static field modulates the energy level of the spin system without inducing transitions. If applied perpendicular to the steady field H_0 transitions may occur whenever

$$\boxed{h\nu = g\beta H_0} \tag{2.6}$$

h being Planck's constant. Commercial spectrometers frequently operate at $\nu = 9{\cdot}5 \times 10^9$ Hz, and for a free electron with $g = 2{\cdot}00$ the magnetic field at resonance is about 3400 G. The instrumentation in electron spin resonance spectrometers may be quite complex, and the reader is referred to a standard text such as the book by Wilmshurst[12]. We do note, however, that because microwave sources can only be tuned over a narrow frequency range it is usual to search for resonances by keeping the frequency constant and varying the magnetic field.

The resonance can be detected only if there is a population difference between the two levels. Although the $+\frac{1}{2} \rightarrow -\frac{1}{2}$ transitions have exactly the same probability of occurring as the $-\frac{1}{2} \rightarrow +\frac{1}{2}$ transitions, the populations in the two levels are governed by Maxwell-Boltzmann statistics, and consequently there are always slightly fewer electrons in the higher level. Thus there will be more transitions from the lower level, tending to equalize the populations. The populations are never equalized because after excitation to the higher level; the electrons relax from the higher state by phonon emission (so-called spin-lattice relaxation) in order to re-establish the Maxwell-Boltzmann distribution. Thus a steady absorption of microwave photons takes place. Since the net magnetization is proportional to the population difference between the states, which is given by

$$\boxed{N_2/N_1 = \exp - (g\beta H/kT)} \tag{2.7}$$

it follows that the magnetization and thus the sensitivity of the system increases with decreasing temperature. A second reason why low temperatures are often essential is that the spin-lattice relaxation time is longer at low temperatures. This has the effect of reducing considerably the width of the absorption line at low temperatures.

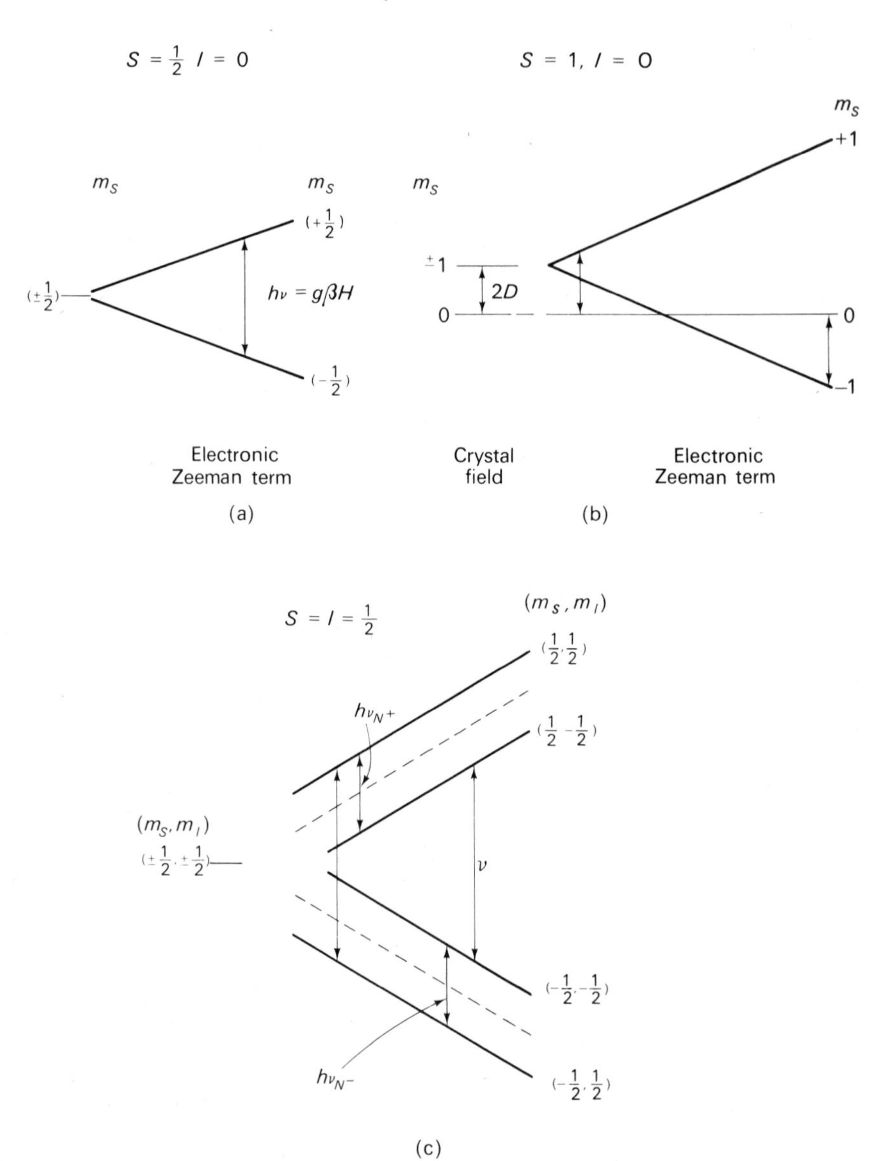

Fig. 2.2 Energy levels and EPR transitions for atoms with (a) $S = \frac{1}{2}, I = 0$ and (b) $S = 1$, $I = 0$. The splitting 2D represents the reduction of spin degeneracy by the internal crystalline field. In (c) the effects of hyperfine splitting in a system with $S = I = \frac{1}{2}$ are considered. The ENDOR transitions are at $h\nu_{N^+}$ and $h\nu_{N^-}$.

The simple doublet state splitting, which produces a single line spectrum, is rarely to be met with in studying defect structure. Indeed it is necessary to consider all possible consequences of the mutual interactions between the electron spin and its environment. The simplest of such interactions is the coupling between the spin and angular momenta. This may arise from the electron's intrinsic orbital angular momentum (i.e., when $L \neq 0$), or when angular momentum is generated in the ground state through overlap of the electron wave function onto the orbitals of neighbouring atoms. As a result the g-value will differ from that for a free electron and, especially in low symmetry crystals, will depend upon the orientation relative to the applied magnetic field. In some cases we are concerned with a system of several electrons with total spin $S > \frac{1}{2}$. The internal electric field of the crystal may then produce a splitting of the levels in the absence of a magnetic field. An example of the splitting of a spin triplet state with $S = 1$ is shown in Fig. 2.2b, where the *fine structure* splitting leads to the $m_S = \pm 1$ and $m_S = 0$ levels being non-degenerate when $H_0 = 0$. The applied magnetic field completely removes the degeneracy and the levels split as shown in the diagram. Since the spin quantum numbers in allowed magnetic dipole transitions obey the selection rule $\Delta m_S = \pm 1$, we observe two transitions in the presence of microwave radiation of the correct frequency.

The final interaction which we consider is that of the electron spin with the nuclear magnetic moments of neighbouring atoms.* This gives rise to hyperfine structure in the spectrum providing that the nuclear spin of the surrounding atoms is non-zero. As an example Fig. 2.2 shows the interaction when the $S = \frac{1}{2}$ spin doublet is coupled to a nuclear spin $I = \frac{1}{2}$. In a magnetic field, there are now four levels characterized by different values of the quantum numbers m_S and m_I. For allowed transitions we have to couple the additional selection rule $\Delta m_I = 0$ to the usual selection rule for magnetic dipole transitions. We recognize that this interaction gives us detailed information of the character and positions of nuclei around a defect. Frequently the information is sufficient to allow a complete identification of the defect.

The hyperfine interactions of defects are often exceedingly complex since they may involve the electron spins interacting with many nuclei. This arises because the electron wave functions spread out all around the centre and may have large amplitudes on nuclei several atom spacings away from the defect. Consequently the electron spin resonance spectrum may consist of large numbers of densely packed and overlapping lines, giving the appearance of a single broad, structureless line. An alternative method of measuring the hyperfine splittings directly is to use the electron nuclear double resonance (ENDOR) technique. To illustrate the principle of ENDOR, we consider a system with total electron and nuclear spins $S = I = \frac{1}{2}$. If the electron spin resonance transitions have long spin-lattice relaxation times, the probability

* In the special case where the defect on which the unpaired electron is located is an impurity atom, an important hyperfine interaction with the impurity nucleus may exist.

of exciting transitions from the upper to lower level is smaller than that for the reverse transitions. Consequently if sufficient microwave power is applied to the sample while the magnetic field is held at the resonant value for the electron spin system, a state may be reached in which the populations of both levels are equal. When this is so absorption of microwave power ceases and the transition is *saturated.* Let us now apply a strong radio frequency pulse and adjust the frequency until at some frequency ν_n, the transitions in Fig. 2.2c between the $m_I = +\frac{1}{2}$ and $m_I = -\frac{1}{2}$ levels are excited. At this point a slight depopulation of the $(\frac{1}{2}, \frac{1}{2})$ level results, removing the saturation condition, so that the electron spin resonance signal between the $(-\frac{1}{2}, \frac{1}{2})$ and $(\frac{1}{2}, \frac{1}{2})$ levels will arise again. In interpreting the ENDOR spectrum it should be realised that the linewidths are related to the nuclear magnetic resonance lines rather than electron spin resonance lines. Since nuclear resonance lines are several orders of magnitude narrower than electron spin resonance lines, the hyperfine interaction constants can be determined with great precision, even when many overlapping electron spin resonance lines are involved. The ENDOR technique also permits analysis of nuclear quadrupole effects, which are seldom easy to measure accurately in electron spin resonance.

2.6 Optical spectroscopic techniques

The optical properties of solids are determined by the way in which the electrons in the material can respond to radiation. Optical techniques have been applied extremely successfully to the study of defects in ionic crystals, and to a lesser extent in semiconductors. Free electrons in metals cause almost total reflection of photons of all energies, and hence studies are restricted to reflectivity measurements. We confine this discussion to phenomena associated with absorption at energy levels in the band gaps of insulators and semiconductors.

Most insulators are transparent to visible light because a large energy gap exists between the valence and conduction bands. Impurity atoms and intrinsic lattice defects with energy levels in the band gap may cause selective absorption of some components of the visible spectrum: the crystals are then coloured. For example sapphire is a transparent ionic crystal consisting of almost pure aluminium oxide. The addition of a few per cent of Cr^{3+} results in broad absorption bands in the blue region of the spectrum, and the characteristic pink colouration of the ruby crystals. Similarly the normally transparent potassium chloride crystals are coloured blue when *F*-centres are present. In both cases the optical transitions give rise to broad bands, the width of which reflects the strength of coupling between the electronic motion and the lattice vibrations. The elemental semiconductors, Si and Ge, exhibit a metallic lustre when viewed in ordinary light. This is because photons with energy greater than about 1 eV excite electrons from states at the top of the valence band into the conduction band, with a consequent absorption of the incident photon. At shorter wavelengths especially in the infra-red these semiconductors become transparent.

Optical absorption and emission measurements are usually made at wavelengths between 1850 Å and 30,000 Å, in which range most commercial monochromators operate. The optical experiments give information about both the defect concentration and the nature of the electronic states involved in the transitions. It is usual in optical absorption experiments to measure the increased absorption of the crystal due to defects as a function of the wavelength of light. The absorption co-efficient, μ, is defined in terms of the radiation intensity, I, and the fractional intensity decrease $-\mathrm{d}I$ per $\mathrm{d}t$ thickness by

$$\mathrm{d}I = -\mu I \, \mathrm{d}t$$

Simple integration yields,

$$\boxed{\mu = \frac{2{\cdot}303}{t} \log_{10} \left(\frac{I_0}{I}\right)} \tag{2.8}$$

Furthermore the integrated absorption co-efficient is related to the oscillator strength, f, of the transition since

$$\int \mu(E) \, \mathrm{d}E = \frac{2\pi^2 e^2 \hbar}{nmc^2} \left(\frac{E_0}{E_{\mathrm{eff}}}\right)^2 Nf \tag{2.9}$$

where N is the volume concentration of defects, E_0 is the average electric field in the medium, E_{eff} is the effective field at the defect, and the other symbols have their usual significance. This equation was first solved by Smakula to give,

$$\boxed{Nf = \text{const.} \times \frac{10^{17} n}{(n^2 + 2)^2} \mu_m W} \tag{2.10}$$

in which n is the refractive index, μ_m the band peak absorption co-efficient, and W is the full band width at half the peak height. The constant is 1·29 for a Lorentzian band shape and 0·87 for a Gaussian band shape. Thus the concentration of centres can be computed from the absorption co-efficient, if the oscillator strength of the transition is known.

Now the process of absorption excites the electron from the ground state into a bound excited level of the defect. The electron may then return to the ground level; in so doing it loses energy to the lattice phonons and by emission of a light quantum. This latter process is known as luminescence. The stimulation of lattice phonons in the transition process results in the bands being very much broader than those customarily observed in atomic spectra. An energy difference between the peaks in the absorption and emission bands is a further consequence of the excitation of lattice modes in the transition process. In view of its importance we now discuss a simple model of the transition process. This model is usually referred to as the configurational co-ordinate model, a model familiar to molecular physicists for several decades.

The physical basis for our model is the adiabatic Born-Oppenheimer approximation: a detailed mathematical treatment is beyond the scope of this book and instead we emphasize the qualitative features introduced by the electron-phonon interaction. In this approximation the eigenstates of the electron moving under the influence of the neighbouring nuclei are calculated by treating the energy of the nuclei as a small perturbation. In view of its small mass the electron responds to the vibrations of the nuclei, whereas the lattice responds only to the average positions of the electrons. The eigenstates of the electron are assumed to be sensitive only to the radial inphase vibrations of the nearest neighbour atoms. Thus the detailed calculation attempts to evaluate the energy of the electron in the ground and excited states as a function of some configurational co-ordinate which specifies the position of the neighbouring nuclei. The vibrations interacting with the ground and excited states respectively have frequencies ν_g and ν_e which are eigenvalues of a harmonic oscillator. Thus the quantum mechanical eigenvalues for the coupled system are $E_g = (n + \frac{1}{2})h\nu_g$ and $E_e = (m + \frac{1}{2})h\nu_e + E_0$, where n and m are the vibrational quantum numbers and E_0 is the energy separation between the states for which $n = m = 0$. These energy levels are shown in Fig. 2.3, where the shape of both ground and excited state curves is parabolic. The discrete quantum mechanical nature of the lattice vibrations limits the allowed values for E_g and E_e to those represented by the horizontal lines in Fig. 2.3 corresponding to the different values of n and m. By comparison with the spectroscopy of atoms where sharp lines are observed, it is evident that the lattice has introduced a large number of discrete energy levels between which transitions may occur. Thus broad bands are to be expected in the optical spectra of solids. Furthermore, since the wave function in the excited state is more diffuse than in the ground state, the minimum energy of the excited state occurs at a different nuclear co-ordinate than in the ground state.

The principal phenomena associated with optical transitions in solids may be described in terms of these configurational co-ordinate diagrams. Consider the case of allowed transitions at 0 K: since only the $n = 0$ vibrational level is occupied, transitions will occur only from this level. Since the time taken for an electronic transition is short compared with the period of atomic vibrations, the lattice co-ordinates do not change during the transition. This is the Franck-Condon principle which leads to the transitions' being represented by the vertical lines in Fig. 2.3. Such transitions occur with varying probability from any position consistent with the spatial extent of the $n = 0$ vibrational wave functions. Thus the transition energy varies as the lattice vibrates between the limits set by the horizontal line representing the vibrational level in the configurational co-ordinate diagram. The probability of any transition occurring from a particular value of the configurational

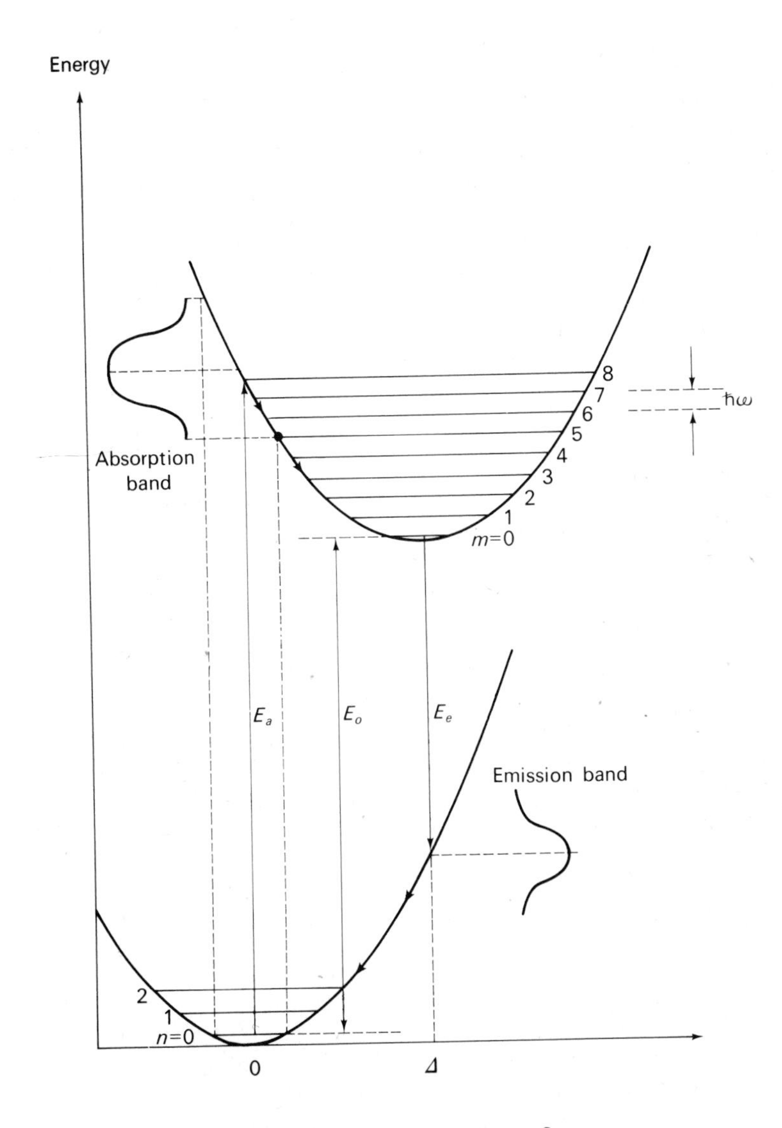

Fig. 2.3 Configurational co-ordinate curves for a defect involving a mean excitation of eight phonons in the excited state. For some defects (e.g. *F*-centres) the excited state relaxation is much greater, typically with $S \gg 20$. The nature of the broad bands at 0K in absorption and emission are indicated. (After HENDERSON and WERTZ. 1968. *Advances in Physics*, **17**, 749.)

co-ordinate is proportional to $|\psi_n(Q)|^2$, where ψ_n is the harmonic oscillator wavefunction for the $n = 0$ state. The maximum value of $|\psi_n(Q)|^2$ occurs at the equilibrium position of $Q = 0$ and transitions from this point represent the peak of the absorption band at 0 K. The width of the peak at this temperature is effectively defined by the spatial limit of the line at $n = 0$. As the temperature increases, different vibrational levels become occupied, which have different probability distribution functions. Detailed consideration shows that the probability function and consequently the band shape is a Gaussian distribution about $Q = 0$. The width of the Gaussian band changes with temperature according as the statistical population of the vibrational states for which $n > 0$. The model predicts that the half-width W_T varies with temperature according as,

$$W_T^2 = W_0^2 \coth (h\nu/2kT) \tag{2.11}$$

Immediately after the excited state configuration has been reached by a transition within the absorption band, the lattice relaxes around the defect to the $m = 0$ vibrational level, by emission of an appropriate number of phonons. Radiative decay from the excited state to the ground state then occurs, the peak position corresponding to the maximum probability in the $m = 0$ state at $Q = \Delta$. Thus since $E_a > E_e$ in Fig. 2.3, it is apparent that there is a shift in the relative positions of the absorption and emission bands: this is usually referred to as the *Stokes shift*.

Figure 2.3 indicates that the greater the number of phonons excited in a transition, the larger will be the Stokes shift. A further consequence is that the weaker the coupling (fewer phonons) the greater is the amount of overlap between the lowest vibrational states $n = m = 0$. If there is an appreciable overlap of the wavefunctions $\psi_n(Q)$ and $\psi_m(Q)$, then we can excite a transition at energy E_0, in which the final $m = 0$ state is reached without the emission of phonons. The probability of a transition between any of the levels is given as

$$P_{nm} = f|\langle\psi_m \mid \psi_n\rangle|^2$$

where f is the oscillator strength of the transition and $\langle\psi_m \mid \psi_n\rangle$ represents the overlap integral between the two vibrational levels. This integral has been evaluated by numerous authors: the general result for a transition involving S-phonons being excited at 0 K is

$$P_{0m} = \frac{e^{-S} S^m}{m!} \tag{2.12}$$

This expression defines the optical lineshape, examples of which are shown in Fig. 2.4. In general when S is small, the probability of observing the zero-phonon transition is greatest. Increasing S effectively increases the

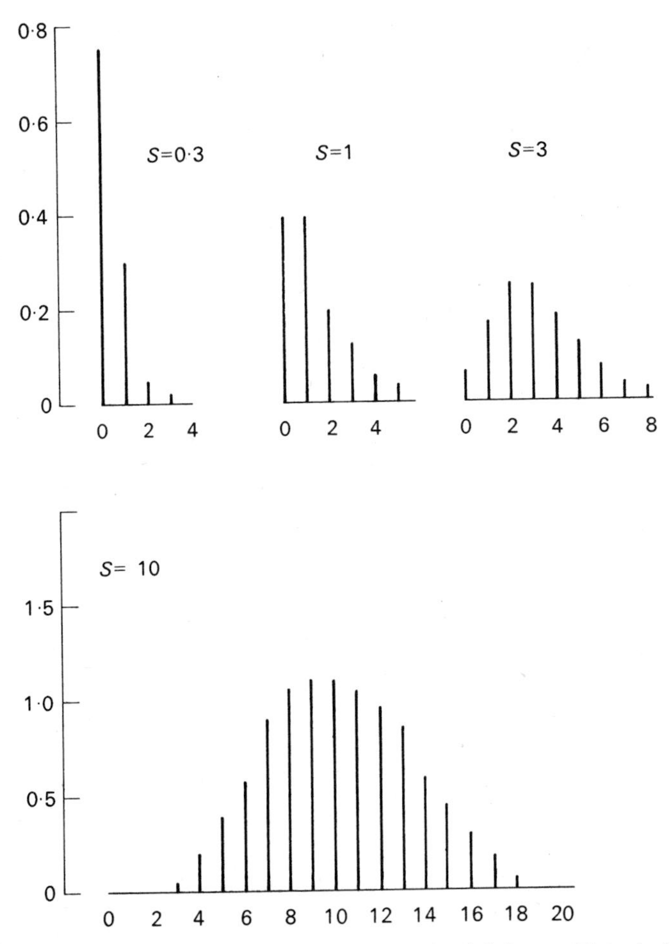

Fig. 2.4 Shape functions for optical absorption bands of defects, with typical values of $S = 0{\cdot}3$, 1, 3 and in the limit of large S where it is evident that the band peak occurs at $S = m$.

intensity in the phonon sidebands at the expense of the zero-phonon line. The sideband structure accompanying the zero-phonon line may give information about the spectrum of lattice vibrations in cases where the defect is coupled mainly to the fundamental lattice modes. When S is very large the shape approaches the Gaussian band, in which the band peak occurs at $S = m$.

Very sharp zero-phonon lines have now been observed in many systems and have been especially useful in investigations of intrinsic defects and impurities in the alkali halides, alkaline earth oxides and some semiconductors. The narrowness of the zero-phonon line allows use to be made of applied external perturbations to investigate the structure of defects. Externally applied elastic stresses, magnetic fields and electric fields have all been used with considerable success.

One further spectroscopic method which should be mentioned is that of photoconductivity. Some defects have excited states near the conduction band and may allow optical transitions in which the electron is excited in the conduction band. Consequently an induced electronic conductivity is observed when the crystal is irradiated with photons of appropriate energy. Studies of such induced photoconductivity can be extremely informative about the nature of the conduction band, the interactions of excitons with defects and the lifetime of the excited levels.

2.7 Internal friction and dielectric loss measurements

Internal friction is observed whenever a body vibrates mechanically at its resonant frequency. In perfectly elastic solids the strain follows the applied stress precisely and the vibrational energy is conserved. Usually the vibrations of the crystal lattice are not perfectly elastic: instead frictional forces of various origins cause dissipaton of the vibrational energy, which is demonstrated by the phase lag between the stress and strain cycles. If the stress and strain vary sinusoidally with time, then we can write $\sigma = \sigma_0 \sin \omega t$ and $\epsilon = \epsilon_0 \sin (\omega t - \gamma)$ respectively, where γ is the phase angle which represents this lag. The internal friction may be measured using either forced or free vibrations. For the latter the logarithmic decrement Δ is related to the phase angle and the quality factor Q^{-1} by

$$\boxed{\gamma = \tan^{-1}(\pi Q^{-1}) = \tan^{-1}\left(\frac{\Delta}{\pi}\right)} \tag{2.13}$$

(The quality factor Q^{-1} is analogous to the quality or Q-factor of a resonant electrical circuit.) For forced vibrations near a resonance the Q-factor is defined by

$$\boxed{Q^{-1} = \Delta\omega/\omega_0} \tag{2.14}$$

where $\Delta\omega$ is the frequency difference between points on either side of the resonant frequency ω_0 at which the amplitude of oscillation is $1/\sqrt{2}$ of the resonant amplitude. The damping of the vibrations or the *anelasticity* occurs as a result of the interaction of the crystal lattice with impurities, intrinsic point defects or dislocation lines.

In this technique the applied stress may cause the defect to jump from one orientation to another, if the defect can achieve a lower energy state by becoming oriented along that direction in which the lattice dimension is changed by the applied stress. If the applied stress varies such that first one

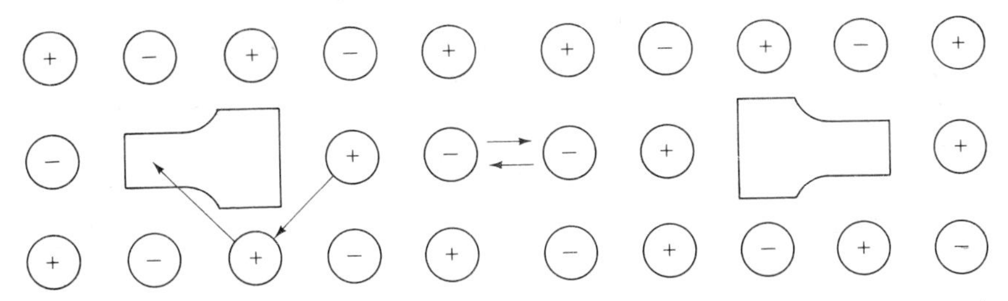

Fig. 2.5 The reorientation of a vacancy pair by cation migration in an alkali halide crystal. An equivalent reorientation may be achieved by anion migration.

direction and then another is stressed in tension at an appropriate frequency, the defect may change its orientation during each cycle, work being done repeatedly. Figure 2.5 shows the reorientation of a Schottky vacancy pair in the alkali halides: the specific mechanism here is by cation motion although anion motion would achieve the same result. In crystals where anion and cation are of similar size these reorientation processes may have similar activation energies, and two relaxation peaks may be observed. Usually the cation moves much more readily than the larger anion and only one relaxation process is observed. Periodically varying stress may similarly cause alternative oscillations of segments of a dislocation line on the atomic places on either side of its original position.

It is usual to investigate the anelasticity as a function of temperature and defect concentration: from these measurements an activation energy E for relaxation between those defect configurations degenerate in the absence of stress can be obtained. The relaxation time, τ, defined as the period for the resonance effect to be produced, then varies with temperature according as

$$\boxed{\tau = \tau_0 \exp (E/kT)} \tag{2.15}$$

Since τ is related to the vibrational period, then the temperature at which the peaks are observed should decrease if the resonant frequency is lowered. The different processes have different characteristic activation energies, which are usually quite small.

An electrical analogue of this mechanical technique measures the dielectric loss of a sample in an applied alternating electric field. This is especially useful when measuring the properties of positive-negative ion vacancy pairs in ionic crystals. Such defect pairs have electric dipole moments, which can give rise to resonant energy loss in an alternating field of frequency equal to that of dipole re-orientation.

2.8 Microstructural investigations

The sizes of defects which may be resolved by microscopic techniques depend partially upon the geometric optics of the microscope system and upon the wavelength of photons used in the microscope. The optical microscope using green light has a limit of resolution not better than 10^4 Å. By using ultra-violet light ($\lambda \simeq 2000$ Å) the limit of resolution can be reduced to about one half the value for green light. Thus optical techniques are especially useful in investigating phenomena associated with solidification, grain boundaries, precipitation, phase transformations and surface topography in solids. Dislocations can only be revealed by special techniques. For example, certain solvents produce well defined etch pits at points of intersection between dislocation lines and crystal surfaces. A further possibility is that dislocations in transparent solids may be revealed by *decoration*, i.e., the precipitation of impurities on the dislocation lines.

X-ray diffraction techniques are usually thought of as means of investigating the structure and lattice parameter of solids, as well as the composition, orientation and degree of perfection. However, recently developed x-ray microscopy techniques may now be used to study individual crystal defects. The method detects differences between the intensity diffracted by regions of the crystal near defects and regions of perfect crystal. Any imperfections give rise to local changes in diffracted or transmitted x-ray intensities and consequently dislocations show up as bands of contrast typically of the order of 5 μ in width. Large areas and crystal thicknesses can be examined using this technique. The method has disadvantages in the low resolution and long exposure times needed. The thickness of crystal that can be studied is an obvious advantage of this technique, which has consequently been used to study many systems including semiconductors, metals and ionic crystals. The features usually revealed by the technique include dislocations, stacking-faults, sub-grain boundaries, grain boundaries and precipitates. They are observed only when these defects have a component of atomic displacement normal to the set of reflecting planes responsible for the diffraction contrast.

The waves associated with moving electrons are diffracted according to the Bragg condition just as are x-rays. The wavelength, λ, of the electrons is given by the de Broglie equation

$$\lambda = \hbar/p$$

where $\hbar$ is Planck's constant and p is the electron momentum. The relativistic correction to this equation amounts to a difference of only 5% to the wavelength of 100 kV electrons. Such an accelerating voltage yields a wavelength of only 0·037 Å. In a thin "perfect" single crystal specimen, the absorption of the electron wave must be expected to be uniform overall and no contrast is observed on the image formed on the zinc sulphide screen of the microscope. Wherever the periodicity of the crystal lattice is sufficiently perturbed by the presence of a defect, the Bragg condition is not obeyed. This *diffraction contrast* can show up the presence of dislocation lines, stacking faults and other defects. Although the resolution of structure down to ~4 Å dimensions is possible in modern electron microscopes, atomic-sized defects such as

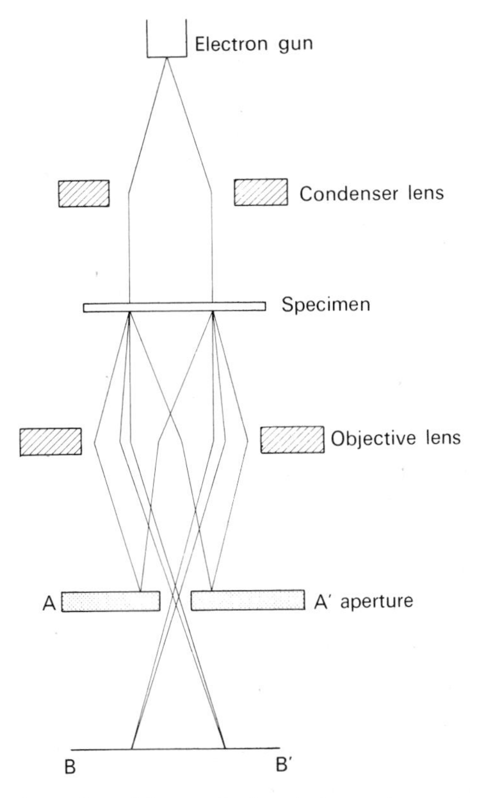

Fig. 2.6 Showing the formation of images in the electron microscope under conditions of bright field illumination. Dark field illumination is achieved by tilting the electron gun until the aperture at AA′ is centred upon a diffracted beam.

vacancies cannot be resolved. Similarly the resolution of the electron microscope does not permit the atomic arrangement around dislocations to be examined, except in some molecular crystals and in Moiré patterns of simple metals.

The similarities in the optical and electron microscopes are demonstrated in Fig. 2.6. The electrons are emitted from a hot tungsten source and accelerated at the anode of the electron gun through 50 to 100 kV. After being focussed onto the specimen in a parallel beam by magnetic condenser lenses, the beam is transmitted through the thin foil and diffracted in a number of directions by the crystal. Each diffracted beam is brought to a focus in the back focal plane of the objective lens at AA′. The resulting diffraction pattern may be suitably magnified and projected onto a fluorescent screen for photographic recording. Apertures of variable size may be inserted at AA′ below the objective lens, whence an image of the lower surface of the specimen is formed at BB′. When the aperture is centred on the main transmitted beam *bright field illumination* is obtained, since then the diffracted beams do not contribute to the image. A perfectly flat, defect-free crystal will then give a homogeneously bright image. Defects in the specimen change the path of the diffracted beam, cause interference with the transmitted beam, and produce concomitant variations in the intensity of the bright field image. The contrast may be reversed to give *dark field illumination* by tilting the electron gun so that the aperture is centred over one of the diffracted beams.

In the neighbourhood of lattice defects, the lattice planes are slightly bent and local diffraction conditions differ from those in the perfect crystal. When the distorted regions of the lattice are in a diffracting position, they change the path of the diffracted beam so that it enters the objective aperture and interferes with the main transmitted beam so changing the transmitted intensity. Thus in bright field illumination a dislocation appears as a dark line of order 50 to 100 Å° wide, much wider than the real width of the dislocation line. The nature of the observed image is actually determined by the precise diffraction conditions being used, and by varying the diffraction conditions we can determine the Burgers vector of the dislocation line. To do this the specimen is tilted relative to the main beam until no contrast associated with the dislocation line is visible. This contrast effect occurs when the planes which contain the Burgers vector, $\boldsymbol{b}$, of the dislocation are not tilted by its presence. The condition for no contrast is written as $\boldsymbol{g} \cdot \boldsymbol{b} = 0$, where $\boldsymbol{g}$ is the reciprocal lattice vector of the reflecting planes. Stacking faults and other defects may also be recognized by the contrast effects which they exhibit in the electron microscope.[13]

The foils used in transmission electron microscopy are typically about 2000 Å thick. Materials with high atomic number absorb electrons most effectively, and must be prepared in thinner sections. In such thin sections both the arrangement and interaction between dislocations may be modified relative to those in bulk samples. This implies some practical limitations on the usefulness of the technique which have been partially overcome by the recent development of electron microscopes operating at much higher accelerating voltages, in the range 1000 to 2000 kV. The additional limitation associated with the inability to reveal the structure of defects of atomic dimensions appears to have been overcome by the field ion microscope, which is able to reveal such microstructural features. In principle this technique is able to show up single vacancies and interstitials, multiple atomic defects, the core structure of dislocations and grain boundaries. Recent measurements on platinum quenched from near the melting point showed that about one lattice site in two thousand was unoccupied. This is consistent with an activation energy for vacancy formation in platinum of 1·2 electron volts.

3

Point defects in ionic solids

3.1 Point defects in thermal equilibrium

In §1.2 we discussed the density of Schottky and Frenkel defects in equilibrium in a monatomic solid. Such calculations when applied to ionic solids take account of the need to maintain charge neutrality. Consider the formation of Schottky defects in the alkali halides: to be general we suppose that there are n_c cation and n_a anion vacancies with formation energies E_c and E_a respectively. The Gibbs free energy of the crystal in the presence of vacancies is then

$$F = E_P + n_c E_c + n_a E_a - TS$$

Neglecting thermal entropy effects, the configurational entropy S is,

$$S = k\left[\ln\left(\frac{N!}{n_c!\,(N+n_c)!}\right) + \ln\left(\frac{N!}{n_a!\,(N+n_a)}\right)\right]$$

At equilibrium the Gibbs free energy is minimized under the constraint that charge neutrality be maintained, i.e.,

$$n_c e_c + n_a e_a = 0 \tag{3.1}$$

In a pure crystal this is satisfied only when $n_c = n_a$: in the more realistic case of an impure crystal the changes δn_a and δn_c are equal as equilibrium is approached. The equilibrium condition is thus satisfied by setting the sum of the partial differentials $\partial F/\partial n_c$ and $\partial F/\partial n_a$ to zero since

$$\delta F = \frac{\partial F}{\partial n_c}\,\delta n_c + \frac{\partial F}{\partial n_a}\,\delta n_a = 0 \tag{3.2}$$

Following the steps outlined in §1.2 we find

$$n_c n_a / N^2 = \exp - (E_c + E_a)/kT$$

In terms of the molar fractions η_c and η_a this becomes,

$$\boxed{\eta_c \eta_a = \exp - (E_S/kT)} \tag{3.3}$$

where E_S is the energy required to form one Schottky defect. Equation (3.3) is written as a *solubility product*, which might also have been derived by applying the Law of Mass Action to the equilibrium. At constant temperature we can rewrite equ. (3.3) as

$$\eta_c \eta_a = \eta_S^2 = \mathrm{K}^{-1} \tag{3.4}$$

This product relationship is valid for both pure and impure crystals. Consequently the concentrations of anion and cation vacancies are related such that when the concentration of one varies, the concentration of the other changes according to Equ. 3.4.

In general, electrical conductivity measurements have given most information about the formation energies, mobilities and association energies of defects in ionic crystals. In these crystals conduction is by transport of charged ions. This process obeys Faraday's laws and is referred to as *ionic* or *electrolytic conductivity*. Many simple and elegant experiments on mass transport in ionic crystals have been reported: they are especially useful since they determine directly the sign of the charge carriers. Much of the accumulated data refers to the alkali and silver halides. There are few well characterized results for the alkaline earth halides and oxides, transition metal oxides and other polar solids of more complex crystal structure. Before considering specific systems the fundamental principles of conductivity methods are outlined by reference to ionic conductivity in the alkali halide crystals.

3.2 Ionic conductivity of pure crystals

Consider the presence of Schottky disorder in an alkali halide crystal at high temperature, where both anion and cation vacancies are mobile. Since their motion is random no current flows in the crystal. When an electric field is applied to the crystal the defects drift in a direction dictated by the applied field and the effective charge on the vacancy. The total conductivity σ is then,

$$\sigma = \sigma_c + \sigma_a$$

σ_c and σ_a being the conductivities associated with the transport of cations and anions respectively. In pure alkali halide crystals, there are equal numbers of anion and cation vacancies. Consequently,

$$\boxed{\sigma = Ne(\eta_c \mu_c + \eta_a \mu_a)} \tag{3.5}$$

where η and μ are respectively the molar fractions and the mobilities of the migrating ions. The quantities μ_c and μ_a are in general different. Substituting for the molar fractions given by Equ. 3.3 into Equ, 3.5 gives,

$$\sigma = Ne(\mu_c + \mu_a) \exp -\left(\frac{E_S}{2kT}\right)$$

In the absence of an applied electric field mass transport occurs by a random process of atoms jumping into their adjacent vacancies. In so doing they overcome a potential barrier. Thus vacancy migration is thermally activated and the mobilities μ_c and μ_a are strongly temperature dependent. The applied electric field lowers the potential barrier in the direction of the field, so favouring ionic migration in this direction. The changes in the potential barrier amount to $\pm e\varepsilon\lambda$, where ε is the applied field and λ is the jump distance. Thus according to Equ. 1.4, the diffusion coefficient in the applied field is

$$D = f\lambda^2 P = f\lambda^2 \exp - \left(\frac{E_M - e\varepsilon\lambda}{kT}\right)$$

where E_M is the activation energy for vacancy migration. This field induced ion transport sets up a concentration gradient $\partial n/\partial x$, defined by Fick's first law of diffusion in terms of the current density, J, by

$$J = D\,\partial n/\partial x \tag{3.6}$$

At equilibrium between the field induced drift current and the opposing diffusion current, the net current is zero, i.e.,

$$ne\mu\varepsilon + eD\,\partial n/\partial x = 0$$

Integrating to obtain the concentration gives

$$n = \text{const} \times \exp - (\varepsilon x\mu/D) \tag{3.7}$$

When ε varies only in the x-direction, i.e., $\varepsilon = -dV/dx$, the Boltzmann equation gives the ion concentration for any value of x as,

$$n = \text{const.} \times \exp - (e\varepsilon x/kT) \tag{3.8}$$

Equating the exponents in Equs. (3.7) and (3.8) yields the Nernst-Einstein equation,

$$\boxed{\mu/D = e/kT} \tag{3.9}$$

Thus the ionic mobility is of the form

$$\mu = \frac{e}{kT} f\lambda^2 \exp - \left(\frac{E_M - e\varepsilon\lambda}{kT}\right)$$

Since $e\varepsilon\lambda \ll kT$ for nearly all practical cases, we can write with reasonable accuracy,

$$\boxed{\mu = \frac{ef\lambda^2}{kT} \exp - \left(\frac{E_M}{kT}\right)} \tag{3.10}$$

Substituting from Equ. 3.10 into 3.5 gives

$$\sigma = \frac{Ne^2\lambda^2}{kT}\left[f_c \exp -\left(\frac{E_M^+ + \frac{1}{2}E_S}{kT}\right) + f_a \exp -\left(\frac{E_M^- + \frac{1}{2}E_S}{kT}\right)\right] \tag{3.11}$$

Equation 3.11 represents *intrinsic conductivity* since it relates the conductivity to the formation and migration energies of intrinsic lattice defects. Actually in most alkali halide crystals $\mu_c \gg \mu_a$ and the second term in Equ. 3.11 can be neglected at all but the highest temperatures. The above treatment is quite general and is easily extended to Frenkel disorder or to mixed Schottky/Frenkel disorder in crystals.

3.3 Impurity conduction in ionic solids

The early measurements on the alkali halides showed two main regions in the conductivity curves where the logarithm of the conductivity is a roughly linear function of the reciprocal temperature. In the high temperature region of intrinsic conductivity measurements are reproducible from specimen to specimen. The low temperature conductivity displays a smaller slope, and depends upon both the particular specimen used and its thermal history. At low temperature we refer to *extrinsic* or *structure-sensitive conductivity*. Impurity ions with the same charge as ions of the host lattice make only a small contribution to conductivity. However, divalent ions in the alkali halides or silver halides, and trivalent ions in the oxides have a pronounced effect. Consider for example the addition of $MnCl_2$ to NaCl. Despite its polyvalent nature manganese dissolves substitutionally on the cation sub-lattice in the divalent state. Each Mn^{2+} ion replaces two Na^+ ions, the second cation site remaining vacant. Thus the electrolytic conductivity is enhanced by there being cation vacancies in excess of the equilibrium concentration. (Precisely the same effect occurs on the anion sub-lattice when CO_3^{2-} or SO_4^{2-} are substituted for the monovalent Cl^- ion.) Assume that a small molar fraction c of the divalent ions is present in the crystal. The total concentration of cation vacancies is now

$$\eta_c = \eta_a + c$$

so that from Equ. 3.4,

$$\eta_S^2 = \eta_c\eta_a = \eta_c(\eta_c - c)$$

which can be solved to give,

$$\eta_c = \frac{c}{2}\left[1 + \left\{1 + \left(\frac{2\eta_S}{c}\right)^2\right\}^{\frac{1}{2}}\right] \tag{3.12}$$

Similarly the concentration of anion vacancies is

$$\eta_a = \frac{c}{2}\left[\left\{1 + \left(\frac{2\eta_S}{c}\right)^2\right\}^{\frac{1}{2}} - 1\right] \tag{3.12a}$$

The form of these equations is shown in Fig. 3.1 for NaCl: the logarithm of the molar fractions of anion and cation vacancies is plotted against temperature for three cation impurity concentrations. Equations 3.12 and Fig. 3.1 clearly express both the change-over from intrinsic conduction at high temperatures to structure-sensitive conduction at low temperatures, and the depression in the anion vacancy concentration caused by the presence of the divalent cations. At high temperature or low impurity concentration Equ. 3.12 becomes

$$\eta_c \approx \eta_s + c/2 \approx \eta_s$$

for $c \ll 2\eta_s$, as expected in the intrinsic range. At low temperature or high impurity concentration ($c \gg 2\eta_s$) Equ. 3.12 becomes $\eta_c \approx c$. Substitution of Equ. 3.12 in Equ. 3.5 gives the conductivity as,

$$\sigma = Ne\eta_s(\mu_c + \mu_a)\left[\left\{1 + \left(\frac{c}{2\eta_s}\right)^2\right\}^{\frac{1}{2}} - \frac{c}{2\eta_s}\left(\frac{\mu_a - \mu_c}{\mu_c + \mu_a}\right)\right]$$

The quantity $Ne\eta_s(\mu_c + \mu_a)$ is the intrinsic conductivity expected for an ideally pure crystal. Thus we obtain

$$\frac{\sigma}{\sigma_0} = \left[\left\{1 + \left(\frac{c}{2\eta_s}\right)^2\right\}^{\frac{1}{2}} - \frac{c}{2\eta_s}\left(\frac{\mu_a - \mu_c}{\mu_a + \mu_c}\right)\right] \tag{3.13}$$

This result has the following general implications: (i) when the impurity concentration is large relative to the concentration of intrinsic defects ($c \gg 2\eta_s$), the ratio σ/σ_0 is linear in concentration since

$$\frac{\sigma}{\sigma_0} = \frac{c\mu_c}{\eta_s(\mu_c - \mu_a)} \tag{3.14}$$

Accordingly when the mobilities are constant at constant temperature isothermal plots of the conductivity ratio versus concentration are linear. This effect is shown schematically in Fig. 3.2a, for addition of the more mobile defect (curve a) and the less mobile defect (curve b).

(ii) When $c \ll \eta_s$ the conductivity isotherms exhibit minima at definite concentrations, which are determined by minimizing $d\sigma/dc$. These minima occur at positive values of c when $\mu_c/\mu_a > 1$, the only case of physical interest. For $\mu_a/\mu_c < 1$ we observe a curved region in the σ/c curve at low concentrations.

(iii) When log σ is plotted as a function of $1/T$ at constant impurity concentration, E_S and E_M may be determined from the slopes of the different regions. This is shown schematically in Fig. 3.2b. The presence of an excess of the less mobile defect depresses the conductivity below that of the pure crystal. An excess of the more mobile defect increases the conductivity

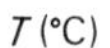

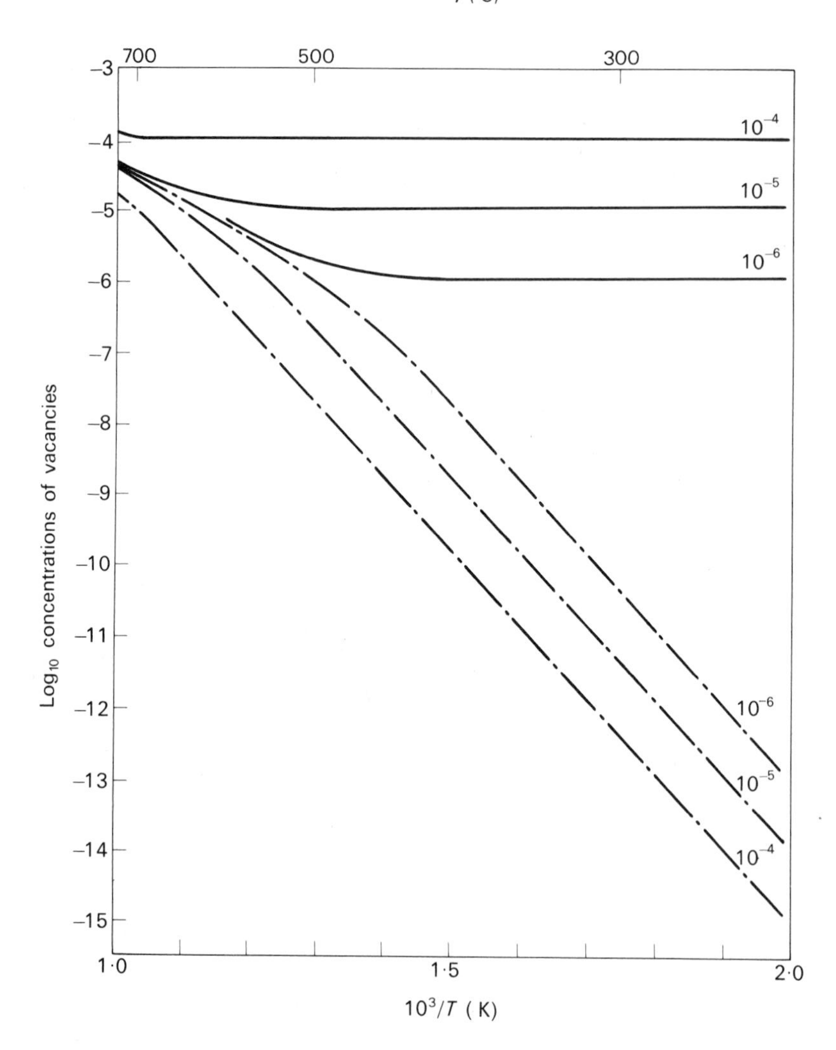

Fig. 3.1 The molar fractions η_c(———) and η_a(–·–·–·) of cation and anion vacancies respectively in NaCl as a function of temperature and divalent cation impurity content as indicated. (After BARR and LIDIARD. 1970. *Physical Chemistry—An Advanced Treatise*. Vol. 10. Ed. Eyring, Academic Press, New York.)

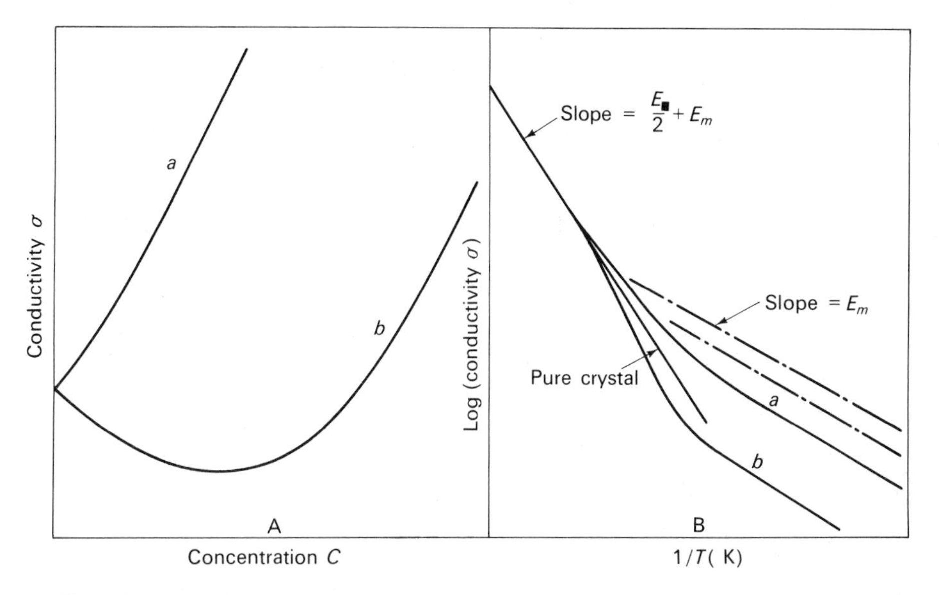

Fig. 3.2 Schematic representation of conductivity of an ionic crystal doped with ions of a different valency as a function of concentrations at fixed temperature (A) and temperature at fixed concentration (B). The curve *a* refers to addition of the more mobile defect and *b* the less mobile defect. (After BARR and LIDIARD. 1969. *Physical Chemistry—An Advanced Treatise*. Vol. 10. Ed. Eyring, Academic Press, New York.)

In Fig. 3.2b the broken curves represent the effects on the conductivity in the extrinsic region of increasing amounts of the more mobile defect.

This discussion has neglected any interaction between defects: but since the cation vacancy bears an effective negative charge it is strongly attracted to a divalent cation. Thus at low temperature we expect an association between these two defects to form a complex. The onset of association between divalent cation impurities and vacancies produces a third stage in the conductivity curve. A consequence of this association is that the conductivity isotherm should be parabolic in form. By fitting the conductivity data to appropriate equations suitably modified from the above discussion, one can determine the cation mobility and the association energy. As the temperature of the crystal is reduced still further, the impurity is precipitated from solution. An activation energy for precipitation may be obtained, although the significance of the result is not clear.

3.4 Diffusion and ionic conductivity

In the foregoing discussion the relationship between mass transport and ionic mobility was demonstrated in deriving the Nernst-Einstein relationship. Accordingly for the transport of cations through the lattice, the mobility is given by

$$\boxed{\mu_c/D_\sigma = e/kT} \tag{3.9}$$

Since $\sigma_c = n_c e \mu_c$, Equ. 3.9 becomes

$$\boxed{\sigma_c/D_\sigma = n_c e^2/kT} \tag{3.15}$$

The diffusion coefficient D_σ calculated from the conductivity measurements can then be compared directly with values of D_σ measured by, for example, the self-diffusion of a radioactive tracer into the crystal. In true self-diffusion the net flow of normal lattice ions through the crystal is measured: these ions are indistinguishable from one another. The radioactive ions are, however, distinguishable from the normal ions and the diffusion coefficient D_T for the tracer experiments differs from the diffusion coefficient D_σ for normal ions. Although D_T is much smaller than D_σ, it is related to D_σ by the concentration n_c/N of diffusing cation vacancies, hence

$$\boxed{D_T = \frac{n_c}{N} D_\sigma} \tag{3.16}$$

Consequently Equ. 3.15 becomes

$$\boxed{\sigma_c/D_T = Ne^2/kT} \tag{3.17}$$

This relation specifically neglects "statistical correlations" or distinguishability effects between the directions of successive jumps of the tracer ion. Such effects arise in the case of the vacancy mechanism in the following way. A tracer ion having jumped into a vacancy still has that vacancy as a neighbour, and thus there is an appreciable probability that the next jump by the tracer ion will be into its former site. This second movement of the tracer ion is not random, it is correlated with the first movement. Consequently we take account of these correlated movements by introducing a factor, f, into Equ. 3.17 which becomes

$$\boxed{\sigma_c/D_T = fNe^2/kT} \tag{3.18}$$

For the vacancy mechanism discussed above the correlation factor should be 1.28. The magnitude of the correlation factor is determined by the atomic symmetry around the particular defect. In CaF_2 where the anion sublattice is simple cubic, the correlation factor, f, is 1·53[14]. Unfortunately the diffusion coefficients involving vacancy migration are rather small and difficult to

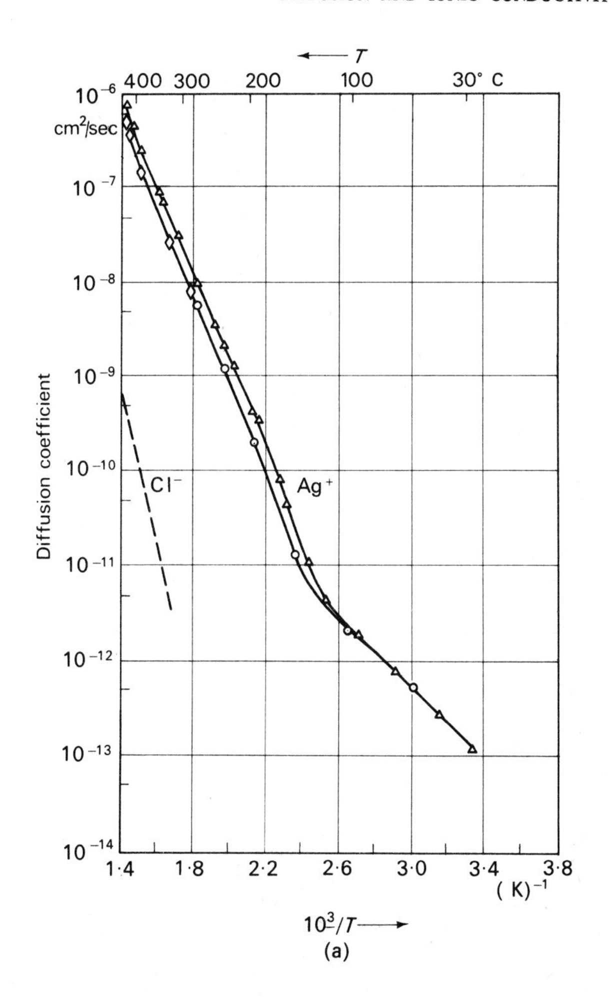

Fig. 3.3 (a) Showing the difference in diffusion coefficient for Ag^+ in AgCl measured by tracer technique ⊗ and ionic conductivity △ which arises as a result of the correlated motion of interstitials. (b) Three possible mechanisms for interstitial displacement in the rocksalt structure. (After MILLER and MAURER, 1958. *J. Phys. Chem. Sol.* **4**, 196).

measure with sufficient precision to test Equ. 3.18. More suitable choices for experimental study are the silver halides, where the diffusion coefficients are high owing to the low defect formation energies.

The silver halides generally show cationic Frenkel disorder, the Ag^+ interstitials occupying the centres of the elementary cubes. The interstitial ions are several times more mobile than the vacancies, and both conductivity and diffusion occur predominantly by the migration of cation interstitials. Figure 3.3a compares the diffusion coefficient of Ag^+ in AgCl determined by

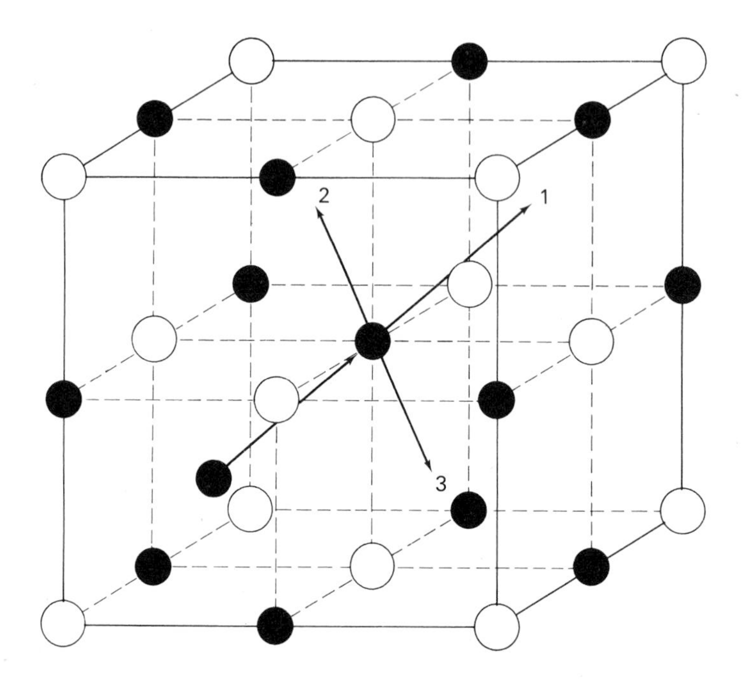

Fig. 3.3 (b)

tracer techniques, with that calculated from the ionic conductivity. The discrepancy at high temperature is due to the correlated motion of the interstitials. The interstitial may migrate by squeezing directly between the atoms on a cube face into a new interstitial site. This cannot explain the differences in Fig. 3.3a since such migration is perfectly random and $f = 1$. The most probable mechanism allows the interstitial ion to jump in a $\langle 111 \rangle$ crystal direction onto the next normal lattice site, the former occupant of that site being pushed into an interstitial site. This can occur in three ways in the rocksalt structure (Fig. 3.3b), one in which the displacements are collinear and two with the displacements making angles of $70°32'$ with each other. Apparently these mechanisms require less energy than that required for diffusion through a (100) face. Including the correlation factors appropriate to the collinear mechanism removes most but not all of the discrepancy in Fig. 3.3a. The possibility of all the above mechanisms operating cannot be ruled out. The major remaining difficulty is to know in what proportions the various possibilities occur.

3.5 Measurements on the alkali halides

The early work in this field is reviewed in the classic papers by Lidiard[14] and by Seitz,[15] and more recent measurements in the paper by Barr and Lidiard.[16] The earlier measurements always revealed the presence of extrinsic conductivity at low temperature. In recent measurements on zone refined KCl Gründig[17] has observed intrinsic conductivity over nine orders of magnitude down to 200°C. A slight change in slope from 1·77 eV to 2·01 eV at high temperature is attributed to an admixture of electronic conduction. A more reasonable explanation of this result might include both anion conduction at high temperature and Debye-Hückel effects which result from interactions between the defects.[14]

As an example of the effects of impurities on ionic conductivity we use the results of a very thorough study of the system NaCl + $MnCl_2$. In this work careful measurements of the D.C. conductivity and A.C. conductivity were made. The D.C. measurements were restricted to low temperatures to avoid electrode effects and space charge polarization. The A.C. measurements were independent of frequency at high temperatures: at temperatures in the association region frequencies near 20 kHz were used to overcome dielectric relaxation associated with the re-orientation of impurity + cation-vacancy complexes. Figure 3.4a represents the conductivity of "pure" NaCl at high temperature: Stage I refers to intrinsic conductivity where cation migration predominates. The slope of this curve corresponds to an activation energy of 1·83 ± 0·01 eV. The activation energy for Stage II, extrinsic conductivity due to unassociated excess cation vacancies, yields $E_M^+ = 0{\cdot}68 \pm 0{\cdot}01$ eV. Since in Stage I the activation energy is $E_M^+ + \frac{1}{2}E_S$, the formation energy for Schottky defects is $E_S = 2{\cdot}30 \pm 0{\cdot}01$ eV. An increased slope at the highest temperatures (Stage I′) is attributed to anion vacancy motion, with an activation energy for anion migration of 2·70 ± 0·20 eV. This interpretation agrees with the anion diffusion data which give E_S^+ in the range 2·12 to 2·70 eV. Despite this apparent consistency with other data it should be noted that Kirk and Pratt made use of the product relationships in Equ. 3.4. These relations specifically omit any long range coulombic interactions between defects, which may result in a non-random distribution of anion and cation vacancies with respect to one another. Thus the interpretation by Kirk and Pratt that Stage I′ is due to anion migration may be unjustified and a further analysis of the data in order.

As noted in §3.3, the log σ versus $1/T$ plot is curved in Stage III (Fig. 3.4b) except in the most heavily-doped crystals. Analysis of this non-linear region and of the conductivity isotherm in the association region indicates that the simple model of nearest-neighbour complexes is applicable for the Mn^{2+} ion-cation vacancy complex. The measured association energy is 0·29 ± 0·03 eV. The precipitation region is also well resolved in doped crystals, and Kirk and Pratt were able to construct a crude solid solubility limit for the NaCl + $MnCl_2$ solid solution.

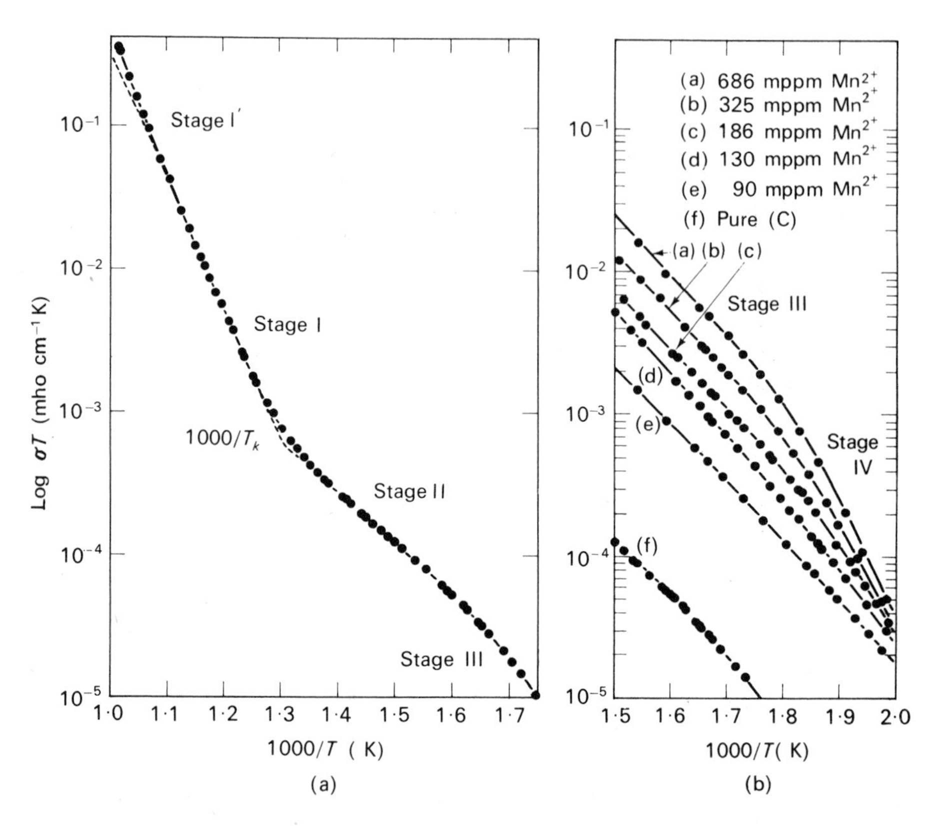

Fig. 3.4 Conductivity measurements of (a) a relatively pure NaCl crystal at high temperature and (b) an NaCl crystal containing various amounts of $MnCl_2$ in the association region. (After KIRK and PRATT. 1967. *Proc. Brit. Ceram. Soc.* **9,** 215.)

Although anion mobility characteristics are sometimes obtained from conductivity measurements, diffusion methods including radioactive tracer techniques are more useful. These methods may isolate the effects of several operative mechanisms. Perhaps the most important result from such studies is that anion diffusion takes place by both a single vacancy mechanism and also by a vacancy-pair mechanism. Since the vacancy pair is a neutral entity, its involvement in anion transport does not contribute to the ionic conductivity. Consequently, the vacancy pair mechanism is not detected in conductivity measurements. Barr and Dawson,[18] investigating the diffusion of K^+ and Br^- in KBr, were able to obtain the activation energies for Schottky defect formation (2·30 ± 0·05 eV), cation mobility (0·67 eV) and anion mobility by the single vacancy mechanism (0·95 ± 0·02 eV) and by a vacancy pair mechanism (2·60 ± 0·03 eV). They also measured the cation vacancy Ca^{2+} ion binding energy (0·59 ± 0·02 eV) and the activation energy for diffusion via dislocation lines (1·49 ± 0·1 eV).

The above discussion is by no means comprehensive, but the systems are representative of the features normally observed in the alkali halides. Although the quantities measured in conductivity and diffusion experiments, e.g. formation and migration energies, are among the most intensively studied parameters in solid state physics, there is still a considerable spread in the experimental results. The range of values for Schottky formation energies is made clear in Table 3.1. In general the lowest values have been observed in more recent studies, presumably as a result of the availability of higher purity crystals.

Table 3.1 A comparison between experimental and theoretical values* for defect energies in some alkali or silver halide crystals

	Formation energy E_s (eV)		Migration energy (eV)				Pair binding energy (eV)
Salt	Expt.	Theory	Expt.		Theory		
	—	—	E_M^+	E_M^-	E_M^+	E_M^-	E_B (Calculated)
LiF	2·2–2·78	1·2–2·7	0·65–0·73	1·1	—	—	0·05
LiBr	1·8	0·9–1·3	0·39	—	—	—	—
NaCl	2·18–2·38	1·3–2·2	0·66–0·76	0·86/2·70	0·70–0·92	1·1/2·54	0·44–0·60
NaBr	1·72	1·3–2·0	0·80	1·18/	—	—	0·37
KCl	2·22–2·3	1·75–2·30	0·71	0·87/2·65	0·75–1·0	1·18/2·43	0·58–0·62
KBr	2·30–2·53	1·7–2·1	0·62–0·67	0·95/2·60	—		0·56
AgBr	1·06	—	0·145/0·350	—	—	—	—
AgCl	1·25	—	0·149/0·384	—	—	—	—

* The values quoted for the alkali halides are taken from numerous sources including reference 16 and lectures by Dr. A. B. Lidiard at A.E.R.E. (Harwell) during 1963–64 and Milan University during 1966, as well as papers presented to the conferences associated with references 18, 19.

Two mechanisms of anion diffusion have been recognized: the lower energy mechanism involves migration via a single anion vacancy and the higher energy mechanism involves migration via a divacancy; they are separated by a solidus in the above table.

For silver halides the formation energy corresponds to anion Frenkel defects, and the migration energies to interstitial mobility (low energy process) and anion vacancy mobility (high energy process).

3.6 Other ionic solids

In the silver halides Frenkel disorder occurs on the cation sublattice, and conductivity measurements in the intrinsic region yield formation energies for cation Frenkel defects of 1·06 eV and 1·25 eV in silver chloride and silver bromide respectively. Since these activation energies are very low, it is obvious that the silver halides have a much higher conductivity than the alkali halides at the same temperature. The Ag^+ interstitials migrate with an activation of 0·145 eV in AgBr and 0·149 in AgCl. The cation vacancy migration energy is approximately $2\frac{1}{2}$ times the interstitial migration energy in both materials. Extensive studies have been made of the extrinsic conductivity in the system AgBr + $CdBr_2$. The presence of Cd^{2+} ions incorporates an excess of cation vacancies in the crystals, which suppresses the interstitial Ag^+ ion concentration. (This is analogous to the suppression of anion vacancies depicted in Fig. 3.1.) As expected this system illustrates quite beautifully the minima in the conductivity isotherms shown schematic ally in Fig. 3.2. A detailed comparison of the results for AgBr + $CdBr_2$ with Equ. 3.13 shows that interstitial to vacancy mobility ratio varies from 7 at 200°C to 2 at 350°C. The mole fraction of defects near the melting point is as high as 10^{-2}, i.e. there is one Frenkel defect per one hundred atoms.

The alkaline earth fluorides and oxides have not yet been intensively studied. In the CaF_2 lattice the usual anion or cation Frenkel defect may exist. However, the Schottky defect consists of one cation vacancy and two anion vacancies. Conductivity measurements suggest that anion Frenkel defects are formed with an activation energy of 2·8 eV, in accord with the theoretical value.[19] The corresponding formation energies for the cation Frenkel and Schottky defects are respectively 7·1 eV/pair and 5·8 eV/trio. Thus Frenkel disorder on the anion lattice should predominate in these materials, although supporting experimental evidence is sparse indeed. The data on oxides such as BeO, MgO and Al_2O_3 are similarly incomplete, all being complicated by impurity problems. Although the formation energies are high, Schottky disorder is expected to predominate.[19]

3.7 Theoretical studies of defect energies

The simple crystal structure and ionic bonding of the alkali halides, together with a wealth of experimental data, have stimulated many theoretical studies of defect energies.[19] The simplest calculation is that proposed by Jost in 1933.[20] The process of forming Schottky defects is divided into two steps:

(i) pairs of ions M^+ and X^- are removed from separated lattice sites inside the crystal to sites on the crystal surface,

(ii) ions neighbouring the vacant sites relax to new positions around the vacancy. In view of the long-range coulomb forces, these relaxations may extend over many concentric shells of ions around the vacancy. In removing an anion or cation from the crystal, energy equal to the lattice energy E_L is expended. In the Born approximation,[14] the lattice energy is

$$E_L = \frac{Ae^2}{R_0}\left(1 - \frac{1}{n}\right) \tag{3.19}$$

Energy equal to $\frac{1}{2}E_L$ is regained when the atom occupies a site on the surface, since the Madelung constant for ions on the surface involves lattice sums over a hemisphere only, the hemisphere outside the crystal being unoccupied by ions. Step (i) consequently requires energy equal to E_L per vacancy, a quantity which can be calculated to high accuracy. The calculation of the energy associated with ions relaxing around the vacancy is much more complex.

Jost[20] represented the vacancy by a spherical cavity of radius R and effective charge e inside a homogeneous material with dielectric constant κ. The dielectric medium is polarized by the charge and an electrostatic potential is set up at the vacancy. If this potential is ϕ_c and ϕ_a respectively for the cation and anion vacancy, the Schottky formation energy is given by

$$E_S = E_L - e(\phi_a + \phi_c) \tag{3.20}$$

Classical dielectric theory gives the polarization as

$$P = \frac{e}{4\pi R^2}\left(1 - \frac{1}{\kappa}\right) \tag{3.21}$$

and the potential at the centre of the vacancy becomes,

$$\phi = \int_R^\infty \frac{D}{r^2} \cdot 4\pi r^2 \, dr = \left(1 - \frac{1}{\kappa}\right)\frac{e}{R} \tag{3.22}$$

Substituting for E_L, ϕ_c and ϕ_A in Equ. 3.20 yields

$$E_S = \frac{Ae^2}{R_0}\left[1 - \frac{1}{n}\right] - \frac{e^2}{2}\left(1 - \frac{1}{\kappa_0}\right)\left(\frac{1}{R_c} + \frac{1}{R_a}\right) \tag{3.23}$$

where R_c and R_a are the effective radii of the cation and anion vacancies. In this idealized model, the radii R_c and R_a are the most difficult quantities to which to assign realistic values. Reasonable values for R_c and R_a are obtained from more accurate calculations of the polarization, which are based upon the actual ionic nature of the crystal.[19] For sodium chloride it is found that $R_c = 0{\cdot}58R_0$ and $R_a = 0{\cdot}95R_0$, and the value of E_S calculated using Equ. 3.23 is 1·86 eV. With the advent of high speed computers, contributions from polarization energies, van der Waals bonding and the elastic energy in the neighbourhood of the distortion around the defect have been

Table 3.2 Experimental and theoretical binding energies between cation vacancies and divalent cations in alkali halides. (After BARR and LIDIARD[16]).

Salt	NaCl		KCl	
Solute	Theory	Experiment*	Theory	Experiment*
Cd	0·38	0·34	0·32	0·5
Ca	0·38	0·31	0·32	0·5
Sr	0·43	0·53	0·38	0·4
Mn		0·29		

* Determined using Stage III conductivity isotherms.

included in the calculations. A comparison of experimental and more recently calculated values of E_S is given in Table 3.1. Good agreement between theory and experiment is observed for LiF, NaCl, NaBr and KCl. The theoretical formation energies are, however, usually too low for lithium salts (except LiF) and theory does not predict $E_S(\text{KBr}) > E_S(\text{KCl})$.

Calculations have also been made of migration energies and binding energies. In all cases, theory confirms the experimental observation that the anion migration energy exceeds the cation migration energy (Table 3.1). Furthermore Table 3.1 demonstrates that experiment and theory agree and that the activation energy for anion diffusion is larger for migration by the vacancy pair mechanism than by the single vacancy mechanism. Although this seems at first sight contrary to expectations, it is a quite reasonable result since the lowering of energy due to the missing cation is more than counterbalanced by the loss of coulomb attraction between the barrier ion and the moving ion. Good agreement is also obtained for the binding energy of the M^{2+} ion-cation vacancy complex (Table 3.2). Perhaps the most satisfying theoretical result was the prediction that in KCl the most stable site for the cation vacancy is the next-nearest neighbour site along the crystal $\langle 100 \rangle$ axis, whereas in NaCl the nearest neighbour $\langle 110 \rangle$ site was favoured. This result has been verified by electron spin resonance for Mn^{2+} impurities.

4

Colour Centres in Ionic Crystals

Colour centre studies have occupied a central position in solid state physics research for several decades, initiated by work done on the alkali halides by Pohl and his colleagues in Göttingen nearly fifty years ago. The term colour centre originally related specifically to intrinsic defects in the alkali halides: today the term embraces all defects which colour insulators, including impurities. The variety of solids now being investigated is emphasized in the penultimate chapter of the book by Schulman and Compton.[21] The present discussion concerns mainly the electronic and atomic structure of intrinsic defects, which have, after years of painstaking work, been unambiguously determined for some materials. Since our understanding is best for defects in the alkali halides, there is an inevitable bias in this discussion towards these crystals. The techniques developed for the alkali halides now provide a foundation for research on other insulators, especially the alkaline earth oxides and halides.

4.1 The structure of colour centres in the alkali halides

In the alkali halides, the band gap between the valence and conduction levels is typically 9 to 10 eV. Photons of appropriate energy liberate electrons from the halide ions, simultaneously producing *positive holes* or simply *holes*. This process corresponds to an electron being removed from the valence band into the conduction band. Less energetic photons do not ionize the anions but instead excite them into higher excited states. These excitations involve transitions of the valence electrons to *exciton states* which produce absorption bands near the fundamental absorption edge of the crystal. The *exciton* is envisaged as a mobile, uncharged particle consisting of an electron in an excited state bound to a positive hole. As indicated schematically in Fig. 4.1, the exciton states lie just below the conduction band.

Vacancies in ionic crystals are effectively charged, and both positive holes and electrons released during irradiation may be trapped in the field of a vacancy of appropriate charge. Electrons are trapped in the anion vacancies to form *F-centres*, which have energy levels within the band gap (Fig. 4.1). The positive holes may similarly be trapped at cation vacancies, if these are

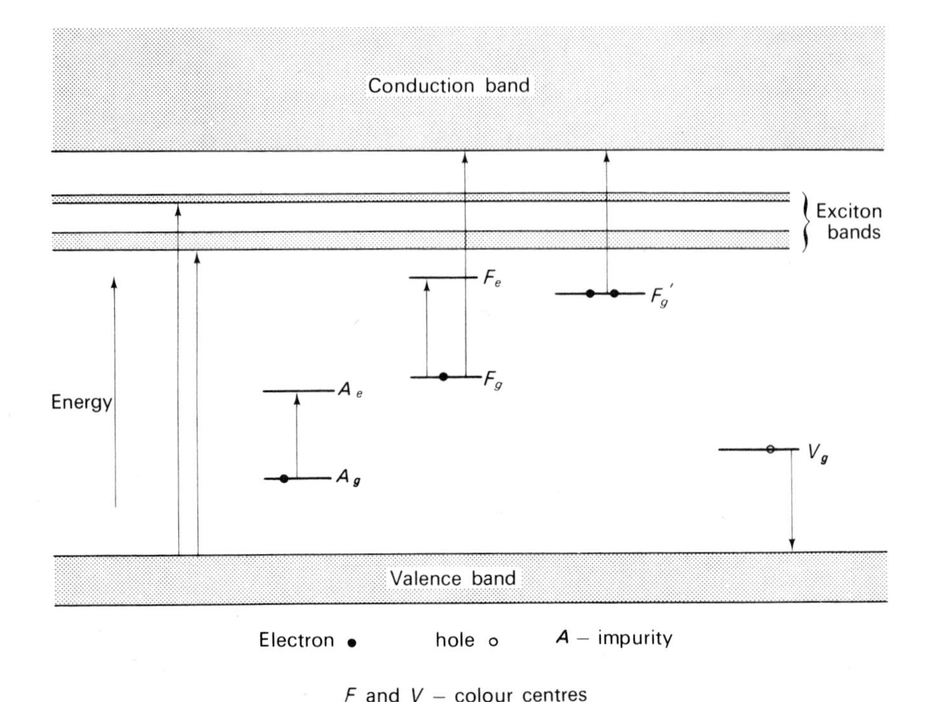

Fig. 4.1 Schematic representation of the band structure of an alkali halide crystal, showing transitions associated with the presence of defect levels in the band gap. The subscripts refer to ground (g) and excited (e) states. (After SCHULMAN and COMPTON. 1962. *Colour Centres in Solids.* Pergamon, Oxford.)

present in the crystal. They may also become *self-trapped* holes or *V centres:* the self-trapping mechanism arises because the neutral chlorine atom is unstable. To overcome this instability the neutral chlorine atom forms a covalent bond along a ⟨110⟩ lattice direction with a Cl^- ion to produce a Cl_2^- molecule ion. These two simple defects and others are shown in Fig. 4.2. Examination of the optical spectrum of an X-irradiated alkali halide crystal reveals a number of optical bands which are usually designated as F, F', F_2, F_3, V_1, V_2 etc. bands.* The structure of the defects shown in Fig. 4.2 have been derived from a study of these bands and other properties. The F'-centre consists of an F-centre at which a further electron

* The nomenclature of defects in insulators developed haphazardly over many years. E. SONDER and W. A. SIBLEY (1972, *Point Defects in Solids*, Ed. Slifkin and Crawford, Plenum, New York) have proposed a nomenclature which is applicable to all defects and all insulating crystals. This system is used here and is explained in the Appendix to this Chapter.

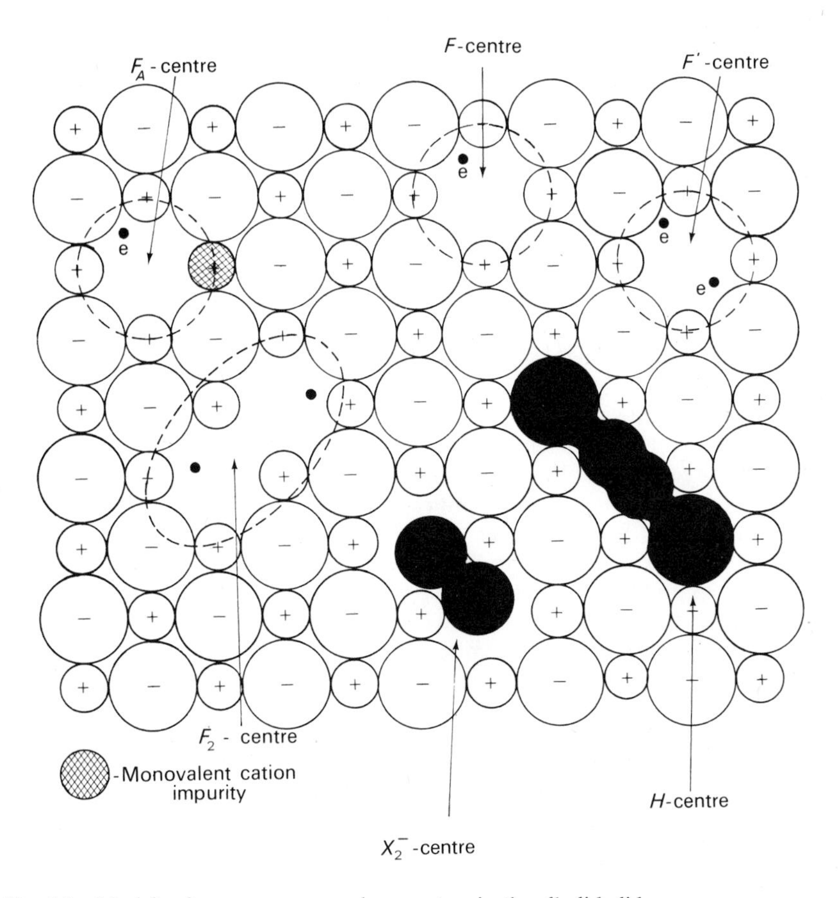

Fig. 4.2 Models of some common colour centres in the alkali halides.

is trapped. Another modified *F*-centre, the F_A-centre results when a neighbouring cation differs from the cations of the host lattice. Other simple defects have the structure of *F*-aggregate centres. The F_2-centre consists of two nearest neighbour *F*-centres along a ⟨110⟩ crystal direction. Addition of a further *F*-centre, to form an array of three nearest neighbour *F*-centres on the (111) plane, is called an F_3-centre. Models for the trapped hole-centres or *V*-centres are less well established. However, a particularly interesting defect, since it resembles the "crowdion" discussed in Chapter 1, is the *H*-centre. In an alkali chloride this defect may be regarded as a Cl_2^- molecule ion located at a single anion site, weakly joined to the two halide ions adjacent to this site. Thus the *H*-centre involves four halogen nuclei distributed over three anion sites.

4.2 The ESR and ENDOR spectra of *F*-centres

The *F*-centre is probably the most widely studied defect, the determination of its detailed structure and properties having occupied a central position in colour centre research. It is unequivocally established that the *F*-centre is an electron trapped at a negative ion vacancy, a model first proposed by de Boer in 1937.[22] Thus the electronic ground level is approximately a (1*s*)-state, the excited states being (2*s*)-like, (2*p*)-like . . . etc. configurations similar to the corresponding states of the hydrogen atom. The hydrogen atom is a spherically symmetric entity in which the electron moves in the field of the central positive charge. The *F*-centre electron moves inside a cage of positive charged associated with the octahedrally disposed cations, occasionally overlapping onto neighbouring atoms. Thus the *F*-centre is more accurately analogous to an "inside-out" hydrogen atom, situated in an octahedrally symmetric environment. Since a single unpaired electron is involved the *F*-centre is paramagnetic, a property which has long been demonstrated by direct susceptibility measurements.

Electron-spin resonance both confirms the de Boer model of the *F*-centre, and discriminates against other models. Investigations of the observed hyperfine structure shows that despite the strong binding of the electron within the vacancy, the electron wave function in the ground state extends out on to several shells of nuclei. Measurements of the number and intensity of the components, the magnitude of the splitting and the angular dependence of the spectra determine the detailed atomic and electronic structure of the centre.

In the alkali halides all the important isotopes possess a nuclear spin, and a complex hyperfine structure is present in all the *F*-centre ESR spectra. In the presence of a static magnetic field H_0, the energy of the unpaired electron spin, E, is written as the sum of the electron and nuclear Zeeman energy and the hyperfine interaction energy, viz:—

$$E = m_s g\beta H_0 - m_I g_N \beta_N H_0 + m_s m_I E_{\text{H.F.S.}} \tag{4.1}$$

where m_s and m_I are the electron and nuclear spin quantum numbers, g and g_N are the electron and nuclear g-values and β and β_N are the Bohr and nuclear magnetons. The hyperfine interaction, $E_{\text{H.F.S.}}$, includes an isotropic term, A, and also an anisotropic term, B, which depends upon the classical dipole-dipole interaction of the electron and nuclear spins. For *F*-centres where the axis of the hyperfine tensor is a simple crystal direction, we may write

$$E_{\text{H.F.S.}} = A + B\,(3\cos^2\theta - 1)$$

Since the nuclear Zeeman energy is small, the transition frequencies are given approximately by,

$$h\nu = g\beta H_0 + m_I[A + B(3\cos^2\theta - 1)] \tag{4.2}$$

Thus for a single nucleus with spin $I > 0$, there are $(2I + 1)$ equally spaced lines. In the presence of more than one nuclei the hyperfine interaction has to be summed over all nuclei with which the electron has a measurable interaction. This is well illustrated by the F^+-centre in magnesium oxide.

Magnesium oxide has the rock salt crystal structure. Thus the anion vacancy has the same symmetry properties as in the alkali halides. When a single electron is trapped at the vacancy, the defect is positively charged and referred to as the F^+-centre. The F^+-centre ESR spectrum is beautifully simple and unambiguous. Each negative ion vacancy has six nearest neighbour Mg^{2+} ions, almost 90% of which are nuclides with zero spin. The remainder are the nuclide ^{25}Mg, which has a nuclear spin $I = 5/2$. The magnetic isotope ^{17}O ($I = 5/2$), is only 0·037% abundant and is not detected. The intensities of the components in the spectrum are calculated from the probability of finding a near neighbour site occupied by a ^{25}Mg nucleus. The probability is determined from the binomial coefficients,

$$P_n(t) = \binom{n}{t} a^{n-t} b^t$$

n being the total number of equivalent sites, t the number of these sites occupied by a magnetic isotope, and a and b the fractional abundances of magnetic and non-magnetic nuclei respectively. Thus a fraction $(0{\cdot}8989)^6$ or 52% of F^+-centres have no magnetic nuclei in the nearest neighbour cation sites. Similarly $6 \times (0{\cdot}8989)^5 \times (0{\cdot}1011)$ or 35·6% of the F^+-centres have one ^{25}Mg-nucleus in the nearest neighbour site and 10% of the F^+-centres have two ^{25}Mg nuclides in the nearest neighbour sites. Other possible combinations are very small.

The ESR spectrum from the F^+-centre in MgO is recorded in Fig. 4.3 with the magnetic field along a $\langle 111 \rangle$ direction. The strong central line at $g = 2{\cdot}0023$ corresponds to the 52% of F^+-centres which have no neighbouring magnetic nuclides. The total intensity of the weaker six line spectrum is 64% of that of the central line and is due to F^+-centres with one cation site occupied by the isotope ^{25}Mg. These lines vary in position according to Equ. 4.21, where θ is the angle between H_0 and the line joining the nearest neighbour site to the centre of the vacancy i.e. a $\langle 100 \rangle$ direction. Thus when the magnetic field is rotated away from the $\langle 111 \rangle$ axis these six lines split into 3 groups, each of six equally intense lines. The values of the constants in Equ. 4.2 for the F^+-centre in MgO are $A = 3{\cdot}94G$ and $B = 0{\cdot}48G$. If the spectrum is examined under high spectrometer gain, (Fig. 4.3b), additional lines from F^+-centres which have two ^{25}Mg neighbours are observed. The hyperfine coupling of these two nuclei is equal and the spectrum is determined by the number of ways of obtaining the different components of the total nuclear spin $I = 5$. For example there is only one way of combining vectorially the two components of nuclear spin $m_I = 5/2$ to give a resultant spin quantum number in the field direction of $m_I = 5$. This will give rise to a

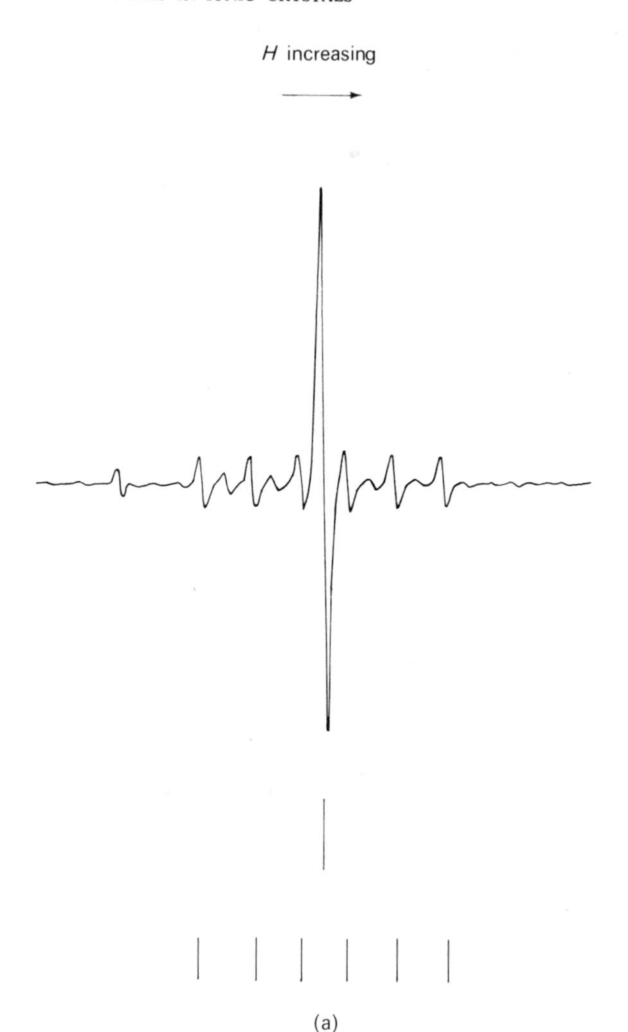

Fig. 4.3 The ESR spectrum of the F^+-centre in MgO, recorded with H_0 parallel to a [111] axis. The nuclear hyperfine interactions between the F^+-centre electron and (a) one ^{25}Mg nucleus as neighbour, (b) two ^{25}Mg nuclei as neighbours, are clearly apparent. (After Henderson and Wertz. 1968. Advances in Physics **17,** 749.)

line of unit intensity. There are two possible configurations of the nuclear spins which yield $m_I = 4$; using an obvious short hand we can write these as (5/2, 3/2) and (3/2, 5/2). Since these two configurations are equivalent they produce a line with twofold intensity. Similar considerations of other configurations shows that the spectrum consists of eleven lines of relative intensity 1:2:3:4:5:6: etc. In the simplest case the two nuclei lie on opposite sides of the vacancy: the splitting pattern is more complex if the two nuclei are not co-axial with the trapped electron except when the magnetic field H lies along one of the four equivalent $\langle 111 \rangle$ axes. In this orientation all the hyperfine lines contribute to a single group of eleven lines with the relative intensities discussed above. The central line corresponds to the configuration $m_I = 0$ and underlies the central line at $g = 2{\cdot}0023$. The other lines are disposed symmetrically about this central line at fields given by $H_0 \pm m_I(A/2g\beta)$ where A is the hyperfine constant for a single nucleus. Thus some

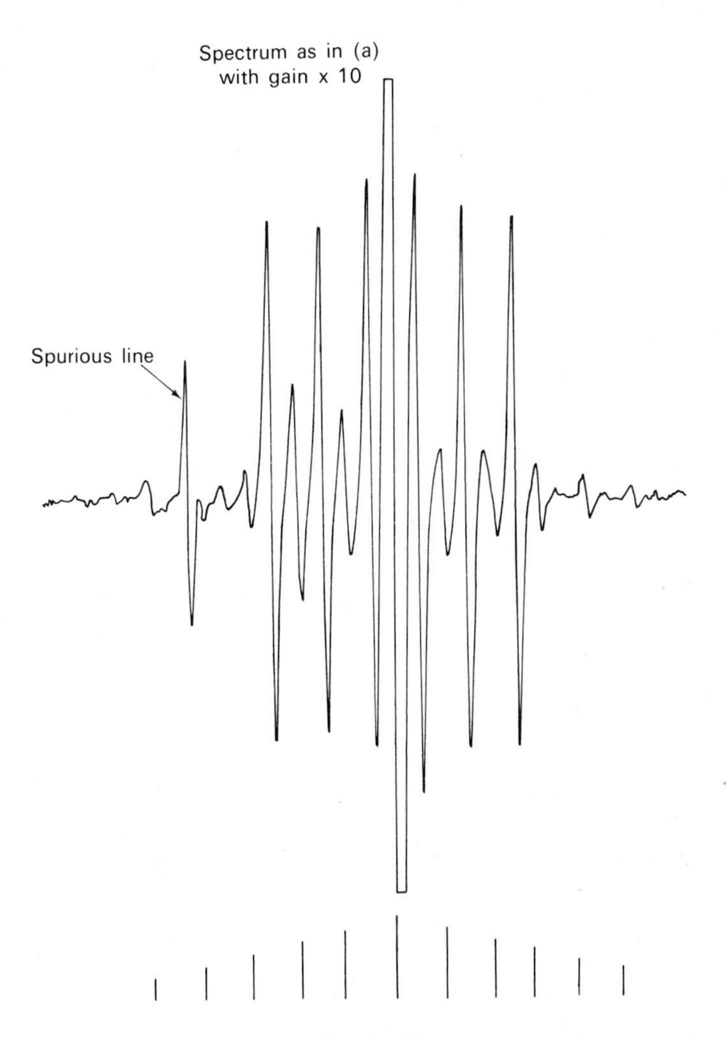

Fig. 4.3 (b)

of these lines lie midway between the six-lines from F^+-centres having one ^{25}Mg nucleus, and are easily recognized in Fig. 4.3b together with the lines which lie outside the more intense central sextet. The intensities of these lines are in accord with predictions based upon the natural abundances.

In alkali halide crystals[23] the F-centre ESR spectrum is much more complex since when there are N-equivalent magnetic nuclei around the F-centre there will be $2NI + 1$ lines. Thus for the nearest neighbour cations in the sodium haldies, where the 100% abundant nucleus ^{23}Na has $I = 3/2$ a spectrum of 19 lines is observed, when H_0 is along a $\langle 111 \rangle$ direction, corresponding to the values of $m_I = 9, 8, \ldots, -9$. The intensities of these lines are given by the statistical weights of the values of m_I, which are in the ratio 1:6:21:56 . . . 580, 546, 456 . . . , 1. Further interactions also occur due to the electron density distribution function extending over more than one shell of nearest neighbour ions. Consequently each of the 19-lines discussed is split into 37 lines in NaCl due to the isotropic interaction with the 12-nearest neighbour Cl nuclei ($I = 3/2$), which constitute the second shell of ions around the F-centre. The spectrum is further complicated by additional splittings from the anisotropic part of the hyperfine interaction, the presence of other magnetic isotopes of the same nucleus, and interaction with higher order concentric shells of nuclei. Thus the ESR spectrum of F-centres in the alkali halides often appear as broad structureless lines. In these cases the hyperfine structure of a particular shell of nuclei is best examined using the ENDOR technique. In interpreting the ENDOR spectrum the selection rule $\Delta m_s = 0, \Delta m_I = \pm 1$ is appropriate, and the ENDOR transitions derived from Equ. 4.2 are given by

$$h\nu_E = -g_N\beta_N H_0 + m_s E_{\text{H.F.S.}}$$

Table 4.1 Parameters which describe the ESR and ENDOR spectra of F-centres in some alkali halide crystals[23] *

			Hyperfine interaction constants (MHz)			
			Shell I		Shell II	
Crystal	g-value	Shells analysed	A/h	B/h	A/h	B/h
LiF	2·0018	I–VIII	39·0	3·2	105·9	14·9
LiCl	2·0018	I–VI	19·1	1·7	11·2	0·9
NaF	2·0001	I–IX	107·0	5·3	96·8	9·8
NaCl	1·9978	I–VI	62·4	3·0	12·5	1·0
KF	1·9964	I–VI	34·3	1·6	35·5	4·1
KCl	1·9958	I–VI	20·7	0·94	6·9	0·52
KBr	1·9829	I–VIII	18·3	0·77	42·8	2·7
KI	1·9649	I–VI	15·1	0·62	49·5	3·0

* Similar results have been obtained in the alkaline earth oxides and fluorides: details are given in HENDERSON and WERTZ, 1968. *Adv. in Phys.* **17**, 749, and STONEHAM, HAYES, SMITH and STOTT, 1968. *Proc. Roy. Soc.* A306, 369.

where ν_E is the appropriate ENDOR frequency. For *F*-centres $S = \frac{1}{2}$ and two ENDOR frequencies are

$$\nu_E = (2h)^{-1}E_{\text{H.F.S.}} \pm \nu_N \tag{4.3}$$

where $\nu_N = g_N\beta_N H_0/h$ is the Larmor frequency of the free nuclei. Thus the nucleus being investigated is recognized by the separation $2\nu_N$ between the pairs of lines in the ENDOR spectrum. Furthermore the dependence of the ENDOR spectrum on crystal orientation identifies the particular shell of the nuclei being investigated since each shell gives a characteristic pattern because of the anisotropic part of the hyperfine interaction. In this way the interaction between the *F*-centre electron and its environment has been measured for many shells of neighbouring nuclei (Table 4.1). In Fig. 4.4 the

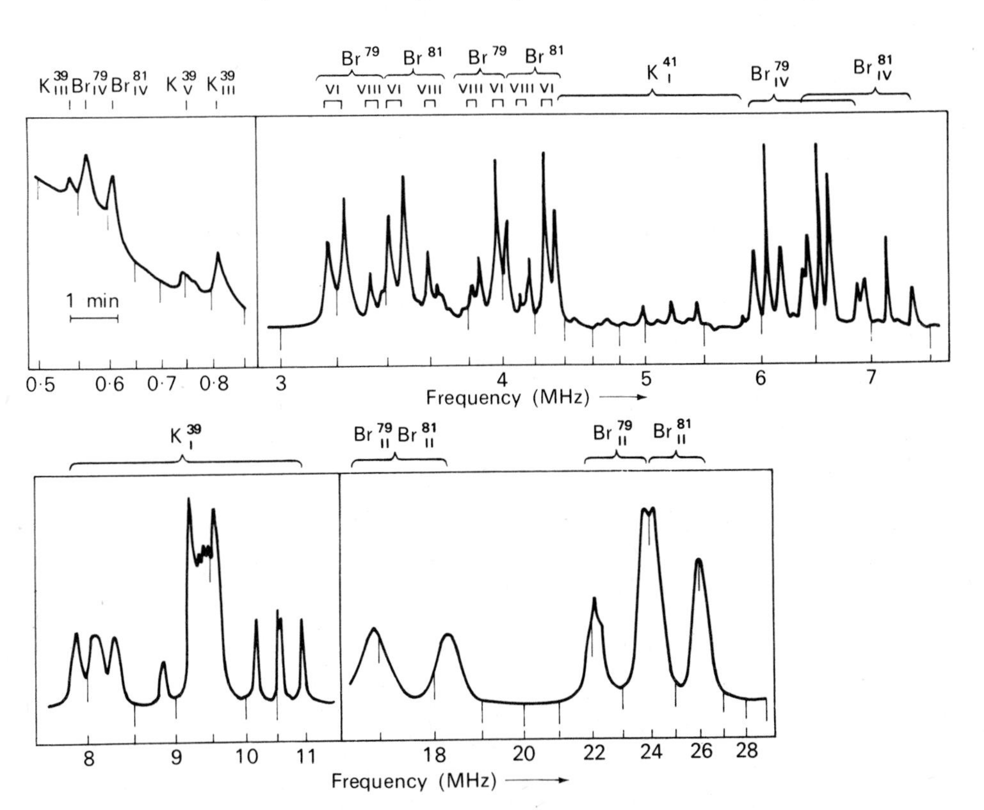

Fig. 4.4 The ENDOR spectrum of *F*-centres in KBr, showing resolved transitions from the nuclei, ^{39}K, ^{41}K, ^{79}Br and ^{81}Br in shells I to VIII. (After SEIDEL and WOLF, 1968. *The Physics of Color Centres*, Ed. Fowler, Academic Press, New York.)

ENDOR spectrum F-centres in KBr shows lines due to the four nuclei ^{39}K, ^{41}K, ^{79}Br and ^{81}Br for shells I to VIII.

The significance of the hyperfine effects in the ESR and ENDOR spectra is that they give a precise measure of the ground state wave function of the F-centre. The isotropic hyperfine parameter, A, is related to the electronic wave function through the Fermi contact expression,

$$A = \frac{8\pi}{3} g\beta g_N \beta_N G\, |\psi(0)|^2 \tag{4.4}$$

G being an amplification factor for the particular nucleus, and $|\psi(0)|$ the amplitude of the electronic wave function at the nucleus. The appropriate values of A for the first shell of ions in all the alkali halides is given in Table 4.1. Evidently since the hyperfine interaction can be measured out to shell eight in some cases, the electron is not as strongly localized inside the vacancy as the simple model implies. It is expected that the value of $|\psi(0)|^2$ will decrease as we go further out from the centre of the vacancy. Typically 63% of the electron density distribution function is inside the anion vacancy, a further 30% inside the second shell and 6% inside the third shell. Less than 1% of the charge distribution is associated with the more remote shells. This is clearly demonstrated in Table 4.2 for shells one to eight around the F centre in KBr: the values $|\psi(0)|^2$ have been normalized relative to the ^{39}K-nuclei in the first shell.

The ground state of the F-centre electron is essentially a $^2S_{\frac{1}{2}}$-state. In free atoms the g-value is given by,

$$g = 1 + \frac{J(J+1) + S(S+1) - L(L+1)}{2J(J+1)}$$

which for spin-only paramagnetism appropriate to a $^2S_{\frac{1}{2}}$-state gives $g = 2{\cdot}00$. Thus the F-centre g-value will approximate that of the free electron. This is confirmed for all the materials referred to in Table 4.1. However

Table 4.2 The ENDOR parameters for shells I to VIII in the KBr F-centre[23]

Shell	Nucleus	A/h (MH$_z$)	B/h (MH$_z$)	$\lvert\psi(0)\rvert^2$
I	^{39}K	18·3	0·74	1
II	^{81}Br	42·8	2·77	0·20
III	^{39}K	0·27	0·022	0·015
IV	^{81}Br	5·70	0·41	0·027
V	^{39}K	0·16	0·02	0·009
VI	^{81}Br	0·83	0·086	0·004
VIII	^{81}Br	0·53	0·06	0·003

there is a significant and easily detected g-shift in almost all cases. This is a further manifestation of the overlap of the F-centre electron onto the orbitals of neighbouring ions. The major part of the g-shift is due to the admixture of ion core p-wavefunction into the wave function of the s-like ground state. In this way orbital angular momentum is generated in the ground state causing deviations from the free spin g-value.

4.3 Optical properties of F-centres

To illustrate the fundamental phenomena which determine the optical properties of the F-centre we treat the trapped electron as a particle in an infinitely deep, three dimensional potential well. Thus the potential energy inside the F-centre is zero and infinite everywhere outside. The time-independent wave function ψ inside the well must satisfy the Schrödinger equation

$$\boxed{\nabla^2\psi + \frac{2mEr}{\hbar^2}\psi = 0} \quad (4.5)$$

under the boundary condition that $\psi = 0$ at $x = y = z = 0$ and a. The normalized eigenfunctions and eigenvalues which are solutions of Equ. 4.5 may be written as

$$\boxed{\begin{aligned}\psi_{lmn} &= \sqrt{8a^{-3}}\sin(l\pi x/a)\sin(m\pi y/a)\sin(n\pi z/a)\\ \text{and}\quad E_{lmn} &= \pi^2\hbar^2(l^2+m^2+n^2)/2ma^2\end{aligned}} \quad (4.6)$$

Consequently the lowest eigenstate of the F-centre with $1 = m = n = 1$ has energy $E = 3\pi^2\hbar^2/2ma^2$. The first excited state is triply degenerate, corresponding to the eigenfunctions ψ_{211}, ψ_{121} and ψ_{112} which have eigenvalues $E = 3\pi^2\hbar^2/ma^2$. Thus the first allowed transition of the F-centre, $1s \rightarrow 2p$, occurs at energy

$$\boxed{E_F = \frac{3\pi^2\hbar^2}{2ma^2}} \quad (4.7)$$

Since the model is crude, this transition energy is not expected to match the experimental results with high precision. A most unsatisfactory feature of the model follows from the boundary conditions, since the probability of finding the electron must be zero everywhere except inside the vacancy: the ESR results discussed above certify that this is totally unreasonable. More realistic models[24] represent the potential energy near the vacancy as a 3-dimensional well of finite potential approximately equal to the Madelung energy. Outside the vacancy the potential energy increases approximately as e^2/r. The reduction in the potential barrier from infinity to a finite value alters the boundary conditions so that the wave-function has a small non-zero value at the edge of the potential well, which drops to zero as $x \rightarrow \infty$. In such cases the electron has a finite probability of being outside the vacancy, although its kinetic energy is less than that required by classical mechanics for scaling the potential barrier.

Despite unattractive features, the particle-in-a-box model is useful since it implies a specific relationship between the transition energy and the lattice spacing of the crystal. Since additional coulombic terms in the potential energy introduce terms in $1/r$ into the Schrödinger equation, we expect E_F to vary as a^{-n}, where n is less than 2. Dawson and Pooley have comprehensively reviewed the dependence of E_F on lattice parameter and conclude that

$$E_F = 57{\cdot}36a^{-1{\cdot}772} \text{ eV} \tag{4.8}$$

Figure 4.5 shows how well the F-band data fit this relationship for all the alkali halides except RbF and LiI. For RbF the F-band transition energy is lower than predicted by Equ. 4.8. Theoretical work attributes this to the electron's extensive penetration in the ground state of the core regions of the large Rb^+ ions. Thus the effective size of the vacancy is larger than expected in the electronic ground state. Consequently E_F is reduced. Although the

Table 4.3 Analysis of the temperature variation of the half-width of F-bands in alkali halides and the peak energies of the absorption and emission bands

				Peak energy (eV)	
Crystal	W_0 (eV)	$E^* = \hbar\omega$ (eV)	S	Absorption	Emission†
LiF	0·688	0·0455	41	5·083	—
LiCl	0·382	0·0207	61	3·256	—
LiBr	0·319	0·0185	54	2·769	—
NaF	0·373	0·0369	18	3·707	1·665
NaCl	0·270	0·0177	42	2·746	0·975
NaBr	0·388	0·0247	44	2·345	—
NaI	0·276	0·0176	44	2·063	—
KF	0·269	0·0149	59	2·873	1·66
KCl	0·195	0·0150	31	2·295	1·215
KBr	0·160	0·0095	51	2·059	0·916
KI	0·146	0·0084	54	1·874	0·827
RbF	0·192	0·0161	26	2·409	1·328
RbCl	0·153	0·0112	34	2·034	1·090
RbBr	0·155	0·0137	23	1·853	0·87
RbI	0·134	0·0089	41	1·706	0·81

* To obtain the effective frequency of the phonon modes coupled to the defect, 1 eV = 2·4181 × 10^8 MHz.

† After Fowler.[24]

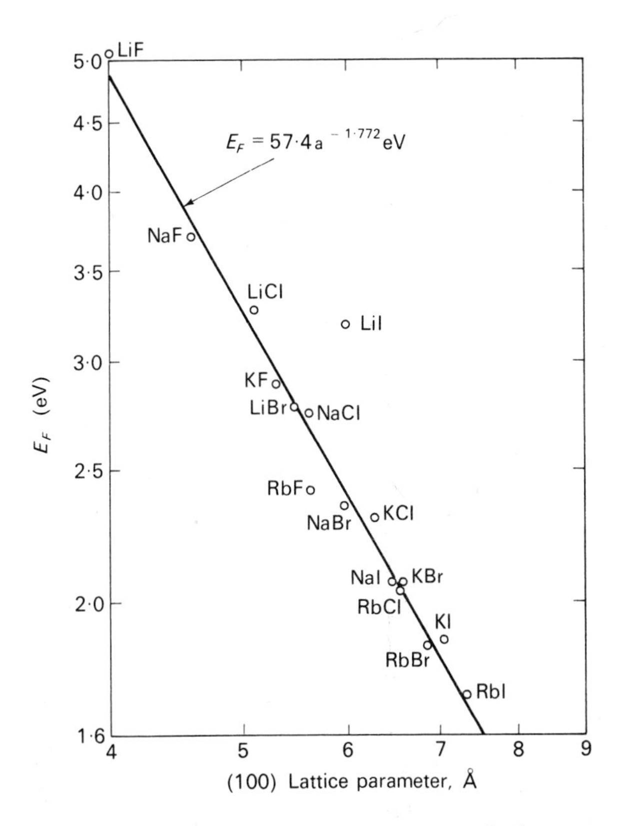

Fig. 4.5 The dependence of the *F*-band transition energy on lattice parameter for alkali halide crystals. (After DAWSON and POOLEY, 1969. *Phys. Stat. Sol.* **35,** 95.)

F-band in LiI has not been unequivocally assigned, the rather large value of E_F in Fig. 4.5 is not unexpected. The penetration of the *F*-centre electron into the large anions is rather less than expected and thus the less extensive ground state wavefunction increases the ground state energy.

The *F*-bands are all broad and almost structureless. The configurational co-ordinate model predicts that the half-width of the band decreases with decreasing temperature according to Equ. 2.11. The usual analysis of the temperature variation of the *F*-band half-width determines $\hbar\omega$ by plotting either W_T against $T^{\frac{1}{2}}$ or $\coth^{-1} (W_T/W_0)^2$ against $1/T$. Both methods give undue weight to particular measurements and the resulting values of ν may be unreliable. A more satisfactory procedure fits the experimental half-widths to Equ. 2.11 using a computerized least squares procedure. The experimental results obey Equ. 2.11 closely and the results are given in Table 4.3, together with some other data for the *F*-centres. Comparison of the frequencies ω with the theoretical phonon spectrum suggests that the phonons responsible for the *F*-band width lie in the acoustic branch for alkali halides with light atomic masses (LiF, NaF, NaCl) and in the optic branch for those with larger atomic masses (KI, KBr). This is a rather broad generalization, however, since the acoustic and optic phonon branches usually overlap. The single frequency model used here does not specify that particular normal lattice mode or localized mode which interacts with the defect. It does imply that the effect of all modes is *qualitatively* well represented by a single mode. For detailed information on the nature of the electron-phonon interaction at the defect more sophisticated techniques must be used. In this respect, recent measurements of Raman scattering from *F*-centres are extremely important since they reflect directly the spectrum of phonons coupled to the defect.

In discussing the optical lineshape in §2.6 we emphasized only the qualitative features of the electron-phonon interaction. Lax[25] has discussed this more fully in terms of the moments (M) of the band shape: for a Gaussian band at T K the second moment is related to the full width at half-height by

$$M_2(T) = H(T)^2/5{\cdot}6 = S(\hbar\omega)^2 \coth(\hbar\omega/2kT) \tag{4.9}$$

and thus from Equ. 2.11,

$$H(0)^2 = 5{\cdot}6\,(\hbar\omega)^2 S \tag{4.10}$$

Applying this result to the alkali halides yields values of S between 18 (NaF) and 120 (LiI). Such large values of S clearly preclude the presence of structure in the absorption spectrum of *F*-centres at low temperature. Furthermore they suggest that the Stokes shift between the peak position in the absorption and emission bands of the *F*-centre should be large. This is found to be the case for all *F*-centres on which emission studies have been reported (see Table 4.3). The position of the emission bands might be expected to follow a similar dependence on lattice constant observed for the absorption bands. Although one can obtain a rather rough correlation, the fit is not nearly as good as for absorption. Figure 4.6 shows the essential characteristics of the emission spectrum from *F*-centres in KBr. The absorption and emission bands are separated by 1·14 eV. As for absorption bands, the emission band broadens and shifts to higher energy with decreasing temperature. Analysis of the temperature dependence of the emission band width yields frequencies which are in general different from that obtained from the absorption results. This result implies that the lattice modes coupled to the *F*-centre ground and excited states are different.

At high temperature, excitation in the *F*-band raises the *F*-centre electron into the conduction band, and the *F*-centre luminescence decreases. Thus alkali halide crystals containing *F*-centres may become photoconducting. Measurement of the temperature dependence, and lifetimes of fluorescence and photoconductivity, yield important information about the nature of the excited 2*p*-state. The model relating the photoconductivity and the *F*-centre

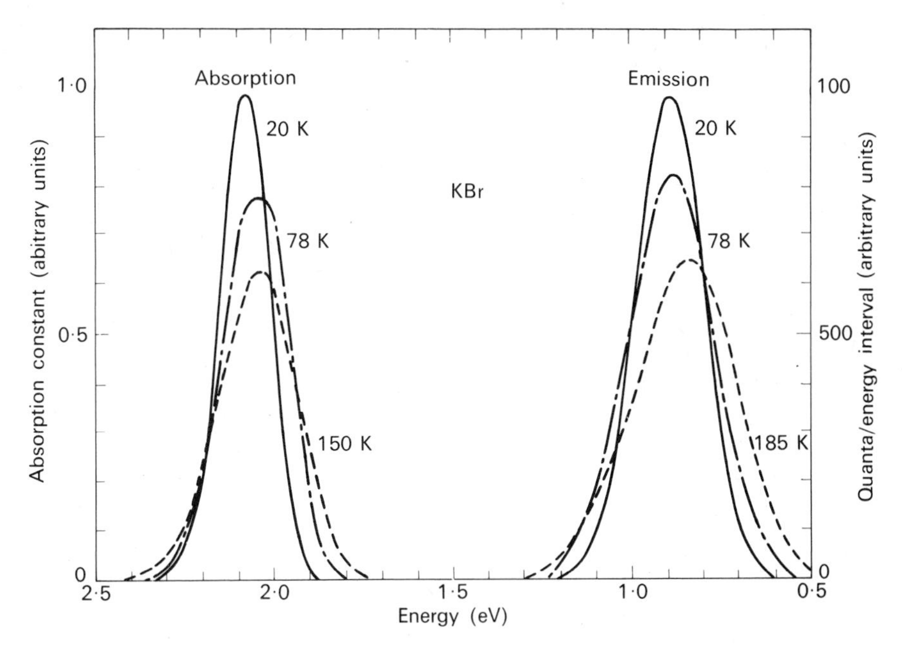

Fig. 4.6 The *F*-centre absorption and emission bands in KBr, as a function of temperature. (After GEBHARDT and KUHNERT. 1964. *Phys. Lett.* **11,** 15.)

fluorescence, shown in Fig. 4.7, assumes that the mean lifetime of the electron in the excited state is determined by several processes: of most importance are radiative decay to the ground state and thermal excitation into the conduction band. The total reciprocal lifetime, $1/\tau$, is written as

$$\frac{1}{\tau} = \frac{1}{\tau_R} + \frac{1}{\tau_i} + \frac{1}{\tau_Q} \tag{4.11}$$

where τ_R, τ_i and τ_Q are respectively the lifetimes for radiative decay, thermal ionization and other processes (including non-radiative decay). The probability of thermal ionization depends upon the Boltzmann factor $\exp(-E_i/kT)$ E_i being the thermal ionization energy of the *F*-centre in its excited state. Hence

$$(1 - \eta_R)\frac{1}{\tau} = \frac{1}{\tau_o}\exp\left(-\frac{E_i}{kT}\right) + \frac{1}{\tau_Q} \tag{4.12}$$

in which $\eta_R = \tau/\tau_R$ is the quantum efficiency for luminescence. Thus measurements of $1/\tau$ and η_R as a function of temperature determine the quantities ΔE, $1/\tau_0$ and $1/\tau_Q$ directly. The comparison of the relative photoconductity and fluorescence of *F*-centres in KCL in Fig. 4.8 emphasizes that these processes are competitive. Swank and Brown[26] determine from these results that the 2*p*-level of the *F*-centre is 0·16 eV below the bottom of the conduction band in KCL, and that the radiative lifetime of the *F*-centre is $0{\cdot}58 \times 10^{-6}\, s$. Apparently this lifetime is too short by a factor of $\sim 10^2$, a property related to the diffuse nature of the 2*p*-state after lattice relaxation.

F'-centres (two electrons in the anion vacancy) are a further important source of photoelectrons. These centres are produced when crystals containing *F*-centres are illuminated in the *F*-band at low temperature; their presence is detected by the simultaneous reduced absorption in the *F*-band and the appearance of a new band, the *F'*-band. Usually the *F'*-band is very broad and overlaps the *F*-band. From studies of the efficiency of producing *F'*-centres by photoconversion from *F*-centres it has become clear that the reaction which takes place may be written,

$$F + h\nu \rightarrow \text{vacancy} + e$$

$$F + e \rightarrow F'$$

Thus the maximum efficiency of *F'*-centre formation should correspond to the destruction of two *F*-centres, as confirmed by experiment. The *F'*-centres are thermally unstable even below room temperature. Since the precursor of the *F'*-centre, the *F*-centre ionization, is thermally assisted it follows that the *F'*-centre can be generated over a very limited temperature range. Both theory and experiment suggest that the only bound state of the *F'*-centre is the

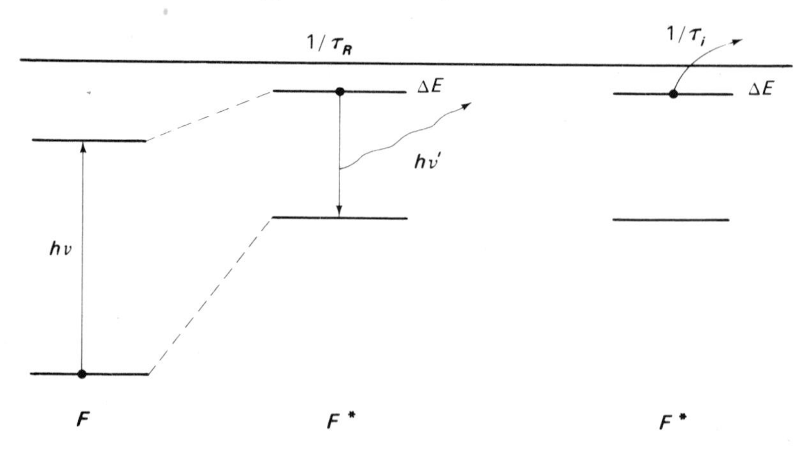

Fig. 4.7 Indicating the absorption ($h\nu$) and emission ($h\nu'$) for *F*-centres, assuming a single excited state. The excited *F*-centre decays by (a) spontaneous emission with radiative lifetime τ_R, (b) by thermal ionization or (c) by other processes. The ionized electron may be trapped at another *F*-centre to form an *F'*-centre. (After Brown, 1967. *The Physics of solids.* Benjamin, New York.)

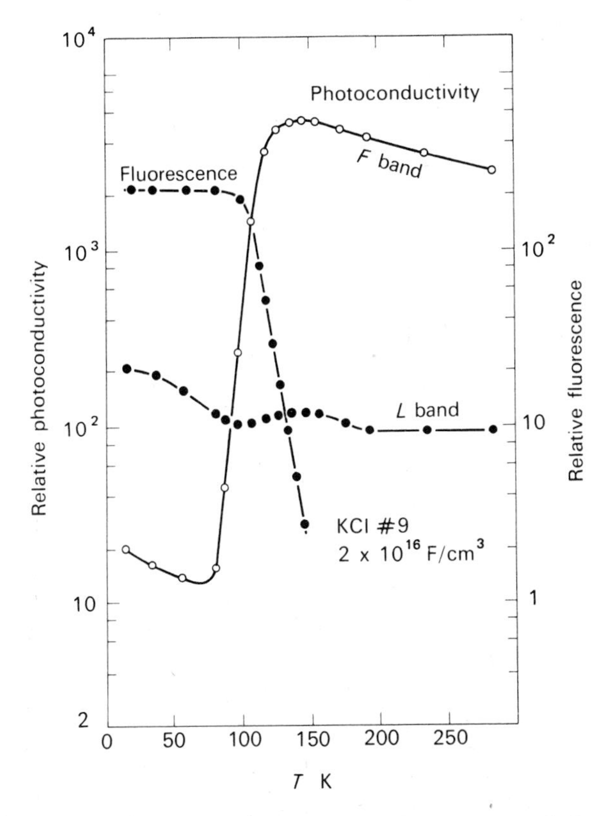

Fig. 4.8 Showing the competing effects of *F*-band fluorescence and photoconductivity. The relative photoconductivity for illumination in the KCl *F*-band freezes out rapidly near 100°K, whereas the *L*-band response is much less temperature dependent. The *F*-centre fluorescence rises rapidly as the photo-response decreases. (After SWANK and BROWN, 1963. *Phys. Rev.* **130,** 34.)

ground state. Thus transitions occur directly into the conduction band, and a broad band corresponding to the ionization of the F'-centre is expected.

4.4 Excited states of the *F*-centre

In the above discussion, the absorption and emission spectra were attributed to transitions between states approximately described as $1s$ and $2p$. By further analogy with the hydrogen atom, we expect transitions $1s \rightarrow np$, where $n \geqslant 3$, to occur at higher energies than the *F*-band.[24] Examination of the absorption spectra of alkali halide crystals containing *F*-centres confirms that several bands, referred to as K, L_1, L_2, L_3, do exist on the short wave length side of the *F*-band. Typically these bands have oscillator strengths,

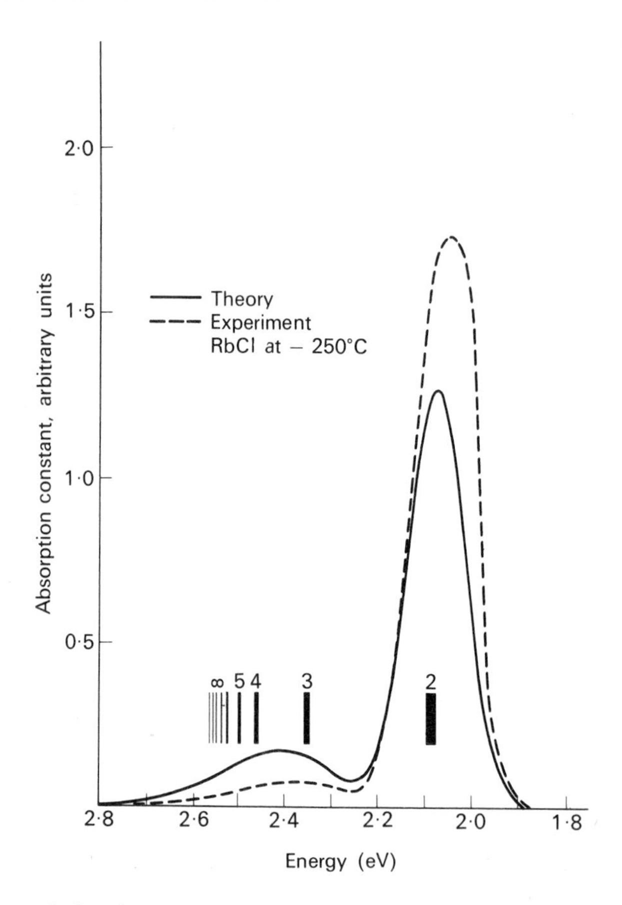

Fig. 4.9 Theoretical and experimental F and K-bands in RbCl. The short vertical lines refer to transitions into $2p$, $3p$, $4p$ etc. states: the width of each indicates its relative absorption strength. The series limit is denoted by ∞. (After SMITH and SPINOLO, 1965. *Phys. Rev.* **140,** A217.)

f, of only 0·1 (K-band) and 0·01 (L-bands) compared with the F-centre for which $f \approx 1$

The K-band is both asymmetric in shape and insensitive to temperature. This is consistent with the K-band transitions being from the F-centre $1s$-level into an infinite number of bound states associated with the coulombic tail of the F-centre potential. A theoretical reconstruction of the F- and K-bands in RbCl (Fig. 4.9) demonstrates how this model predicts the essential asymmetry of the K-band. F-centre transitions into the conduction states begin in the high energy tail of the K-band. This is demonstrated by observation at 4·2 K of a rapid increase in the photoconductive response as the photon energy is increased within the high energy tail of the K-band. The quantum yield for photoconductivity is even higher in the region of the *L*-bands. Consequently, it follows that the *L*-bands are transitions from the *F*-centre ground state to quasi-discrete states degenerate with the conduction levels. The fluorescent emission from *F*-centres is the same whether excitation takes place in the *F*, *K* or *L*-bands.

4.5 Perturbed *F*-centres

The optical properties of *F*-centres are normally complete isotropic. They may become anisotropic when externally applied perturbations, such as a uniaxial stress or an electric field, slightly reduce the *F*-centre symmetry. Studies of such effects have yielded much information not normally obtainable from the unperturbed *F*-centre. A strong internal perturbation is present in the F_A-centre, where one of the six nearest neighbour cations is replaced by a different alkali ion (Fig. 4.2). Thus the usual octahedral symmetry (O_h) of the *F*-centre is reduced locally to tetragonal symmetry (C_{4v}). The $\langle 100 \rangle$ symmetry of F_A-centres was demonstrated by optical and photochemical work, while the atomic model was verified using ENDOR. We discuss briefly some of the optical propertes of the F_A-centre; a comprehensive treatise on these properties is given by Lüty.[27]

F_A-centres are formed in impure crystals containing alkali metal differing from the host lattice (e.g. Li in KCl or Na in KCl). Crystals are irradiated with *F*-band light at temperatures suitable for the photoconversion

$$2F + h\nu \rightarrow F' + \text{anion vacancy}$$

to take place. If the crystal is held at this temperature for some time in the dark F_A-centres are formed. The F_A-centre concentration is always proportional to the concentrations of both cation impurites and to the primary products of photoconversion, the F' centres/anion vacancy pairs. Thus the absorption of light does not itself produce the F_A-centres: it does produce the F'-centre/anion vacancy pairs to which the cation impurities diffuse during prolonged storage at the photoconversion temperature. Studies of the kinetics of F_A-centre formation give important information about the low temperature diffusion of impurities in the alkali halides. For example when F_A(Na) centres are produced in KCl, Na^+ ions diffuse with an activation energy of only 0·60 eV, much lower than is observed for radioactive tracer diffusion.[27]

The consequences of locally reduced symmetry around the F_A-centre are shown in Fig. 4.10: since the 1*s* ground state is spherically symmetric it is almost unaffected by the perturbing influence of the impurity ion. Inspection of Fig. 4.10a, where the angular form of the 2*p*-wave functions are shown relative to the positions of the six neighbouring cations, reveals that the $2p_z$ energy level is lowered relative to the $2p_x$ and $2p_y$ levels. Thus the three fold degeneracy of the *p*-state is removed and two levels are formed one a singlet the other a doublet. Thus in Fig. 4.10b there are two transitions,

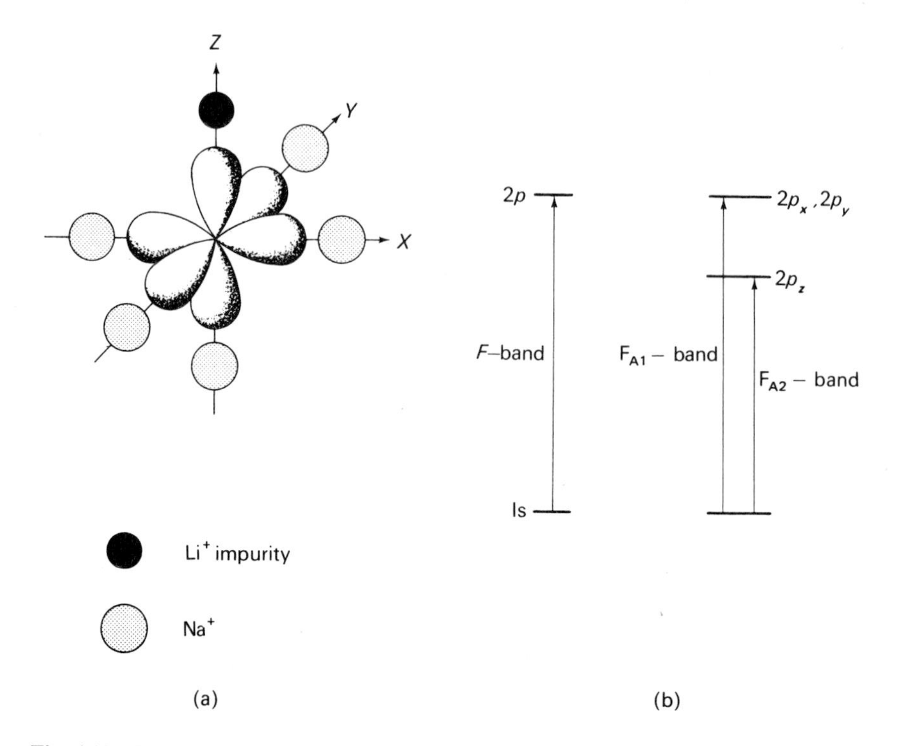

Fig. 4.10 Showing (a) a schematic representation of the environment of F_A-centre in the excited $2p$-level for the system NaCl-Li, (b) the splitting of the sp-levels by the axially symmetric distortion consequent upon the Li^+ ion replacing one of the six octahedrally disposed Na^+ ions around the vacancy.

F_{A2} being to the degenerate p_x and p_y levels is almost co-incident with the F-band, whilst the F_{A1}-band occurs at a lower energy. Since the transitions correspond to charge oscillations along the directions of the p_x, p_y and p_z-states, the two components of the optical absorption spectrum of the F_A-centres are plane polarized, F_{A1} in the direction of the impurity and F_{A2} in the plane perpendicular to the impurity.

A further directional effect may also be produced in the absorption spectrum. When a crystal containing F_A-centres is irradiated at a sufficiently high temperature with light polarized appropriately for the F_{A1} or F_{A2} transitions, an ionic reorientation takes place which produces an alignment of the tetragonal axes of the defect. Thus instead of a third of the defects each having their C_4-axis parallel to one of the ⟨100⟩ directions, more and more of the defects adopt an orientation in which they are no longer excited by the incident polarized light. An example of the dichroism induced by optical alignment of F_A-centres in KCl is shown in Fig. 4.11. Reorientation consequent upon irradiation with F_{A2}-band light polarized along the [100] axis, (which excites F_{A2}-transitions for the (010) and (001) oriented defects only), decreases the number of centres aligned along the [010] and [001] axes. Measurement with [100] polarized light reveals a decreased absorption in the F_{A2} band and an increased absorption in the F_{A1}-band. A corresponding increase in the F_{A2}-band and decrease in the F_{A1}-band is observed with measuring light polarized in the [010] direction. The defect reorientation is thermally activated and may be frozen in at sufficiently low temperatures. Studies of the optical reorientation as a function of temperature determine the activation energy for migration of the impurity ion to an equivalent near

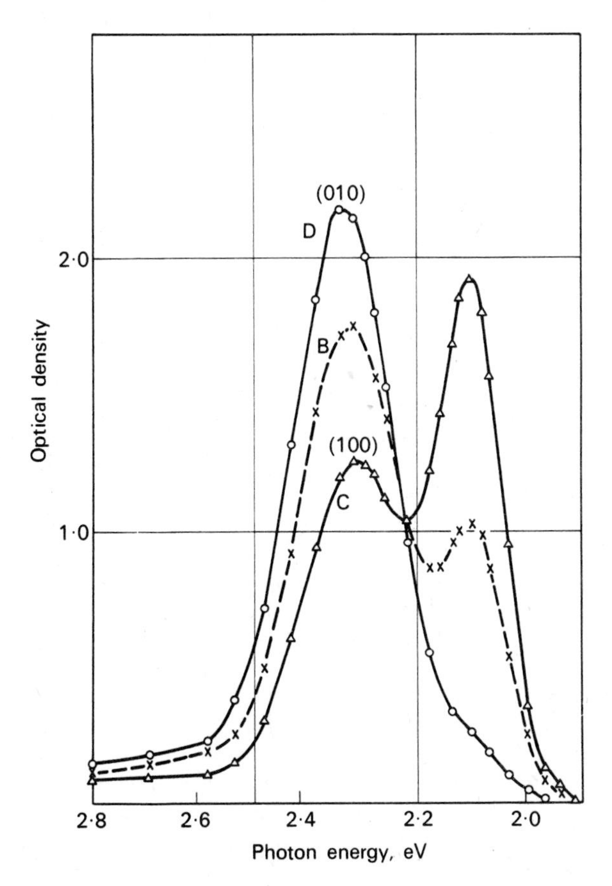

Fig. 4.11 Optical absorption spectrum of F_A-centres in additively coloured KCl containing initially $6{\cdot}8 \times 10^{16}$ cm^{-3} *F*-centres. Spectrum B was obtained by bleaching with *F*-band light at -30°C followed by an optical bleach of the *F*-band. The crystal was then irradiated at -180°C with [100] polarized *F*-band light. Measurement with [100] polarized light gave spectrum C and with [010] polarized light spectrum D. (After SCHULMAN and COMPTON. 1962. *Colour Centres in Solids.* Pergamon, Oxford.)

neighbour site of the defect. Typical activation energies in the range 0·02 to 0·11 eV are observed. Reorientation consequent upon thermal annealing occurs with activation energies near 1·0 eV. This energy difference is due to optical alignment taking place *after* excitation and in the *relaxed* excited state: the electronic state of the defect is seen to have an astonishingly large effect on the activation energy for diffusive movements.

The dichroic absorption spectrum of the F_A-centres induced by defect reorientation facilitates the analysis of the properties of the F_{A1}- and F_{A2}-bands along the lines outlined for F-centres. Both the F_{A1}- and F_{A2}-band shift to lower energies and broaden with increasing temperature, the phonon frequencies broadening the transitions being approximately equal, differing very little from that of the F-centre. The close similarity of the F_A-centre and F-centre in absorption is confirmed by the two transitions $1s \rightarrow 2p_z$ and $1s \rightarrow 2p_x$ and $2p_y$ having i) the same oscillator strengths and ii) the same integrated absorption as the F-centres from which they were formed. Furthermore at higher energies than the F_{A1}-band weaker absorption bands K, L_1, L_2, etc. are observed.

Despite there being two absorption bands due to the F_A-centre only one emission band is observed, whether excitation occurs in the F_{A1}-band or F_{A2}-band. In §2.6 it was shown that emission takes place only after the lattice has relaxed from the configuration reached in the absorption process to a new configuration of minimum energy in the excited state. In general the two configurational co-ordinate curves for the F_A-centre excited state should result in there being two emission bands which are Stokes-shifted relative to the absorption bands. The observation of a single emission band suggests that a radiationless transition might occur between the higher and lower level. This lower level is characteristic of the $2p_z$-state and the luminescence should be strongly polarized. In fact the emission is almost completely unpolarized, and a different mechanism is required. A likely explanation is that during absorption the electron reaches a very compact excited state which feels strongly and is split by the effects of lower symmetry. When the lattice relaxes, a very diffuse state exists which does not feel the effects of the perturbing impurity. For this relaxed excited state the p-state splitting is removed and one unpolarized luminescent transition is observed. This emission has quite similar properties (for example, peak position, band width, lifetime) to the F-centre emission. F_A-centres in the systems KCl:Li, RbCl:Li and RbBr:Li have quite different characteristic emission spectra. In these systems excitation in the F_{A1} or F_{A2}-bands produces one emission band which has both an unusually large Stokes shift and a half-width reduced by a factor of almost five from the F-band emission.[27]

4.6 *F*-aggregate centres

A number of absorption bands are observed in the optical spectra of the alkali halides which are assigned to transitions at defects consisting of clusters or aggregates of *F*-centres. These defects are created when crystals containing *F*-centres are irradiated with *F*-band light. The F_2-bands are the first products of this treatment, followed by the F_3-bands; in both cases the bands are observed at lower photon energies than the *F*-band. Since the *F*-band decreases as the F_2- and F_3-bands increase, *F*-centres are evidently a necessary prerequisite to the formation of the aggregate centres. By analogy with the earlier discussion on the *F*-centre, the electronic states of the F_2- and F_3-centres are regarded as similar to those of the H_2 and H_3 molecules. The

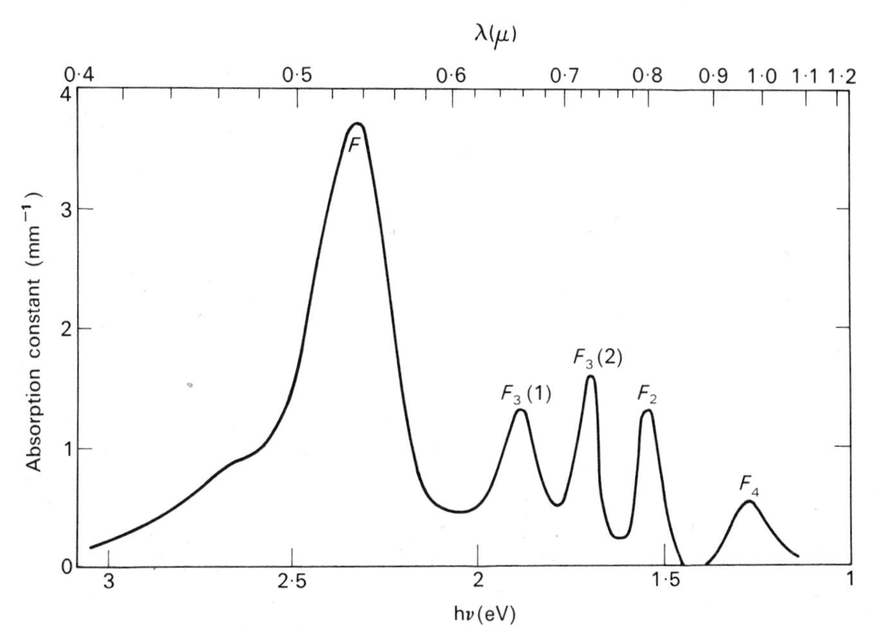

Fig. 4.12 Optical absorption due to various electron-excess colour centres. (After VAN DOORN. 1962. *Philips Research Reports.* Supp 4.)

transitions between the appropriate states are of course modified by both phonon coupling and by the removal of electronic degeneracy by crystal field effects. Figure 4.12 shows the spectral position of the aggregate bands relative to the *F*-band in KCl; other transitions also occur but underlie the *F*-band.

The F_2-centre

The experimental evidence must demonstrate that there are two *F*-centres in nearest neighbour sites along a [110] crystal axis, so that the overall symmetry is orthorhombic (D_{2h}). The optical transitions may be qualitatively thought of as between the (1*s*) ground state of two *F*-centres and excited states which are products of (1s)-functions at one site and (2p)-functions at the other site. Transitions excited by light polarized perpendicular to the defect molecular axis (z-axis) are closely similar in energy to the F-centre transitions since only $2p_x$ and $2p_y$ states are involved. Since they are overlapped by the F-band they are not well understood. The principal F_2-band is excited by light polarized parallel to the z-axis of the defect and Fig. 4.12 shows this band to be easily separated from other bands in the spectrum. The detailed quantum mechanics of the hydrogen molecule determines that the electronic ground state is a spin singlet and that the excited states are either spin singlets or spin triplets. Since transitions between spin singlets and spin triplets are forbidden,[28] the strong optical transitions of the F_2-centre are necessarily between the singlet ground state of the defect to excited singlet states. For F_2-bands in the alkali halides, Ivey[29] derived an empirical relationship between the peak energy of the band, E, and the lattice constant of the crystal d of the form,

$$E = kd^{-1\cdot 56} \text{ eV}$$

A particle-in-a-box model of the F_2-centre predicts an expression in d^{-2}. However the strong electron-electron interaction along the z-axis introduces terms in d^{-1} into the expression for the energy, so that the experimental dependence on $d^{-1\cdot 56}$ is not unexpected.

There are six equivalent orientations of the F_2-centre in the rocksalt crystal structure, each being equally populated with centres. Thus the F_2-band absorption will be isotropic, irrespective of the polarization of the light or the orientation of the crystal relative to the incident light. An unequal distribution of centres among the equivalent orientations results in dichroic F_2-band absorption similar to that discussed earlier for F_A-centres. The F_2-centres in particular orientations may be optically bleached near room temperature using appropriately polarized light in the F-band. The number of centres bleached by the incident photons will be proportional to the intensity of light absorbed by the centres. Thus if we illuminate the crystal with light perpendicular to a (100) crystal face and polarized along the $\langle 011 \rangle$ direction, only centres aligned in the [011], [110], [$1\bar{1}0$], [101] and [$10\bar{1}$] axes will absorb light, since the absorbed intensity is proportional to $\cos^2 \theta$, where θ is the angle between the direction of polarization and the electric dipole oscillators. The rates of bleaching of these centres are in the ratio 4:1 for centres aligned along [011] relative to centres aligned along the other affected axes. Centres aligned along [$0\bar{1}1$] do not absorb and hence are not bleached. Consequently the F_2-band absorption for measuring light polarized in the $\langle 110 \rangle$ and $\langle 1\bar{1}0 \rangle$ directions will be different. Thus study of the dichroism for a number of different orientations and polarizations of the bleaching and measuring light allows one to define the optic axis of the defect. Ueta[30] has found excellent quantitative agreement between the experimentally observed dichroism of the F_2-centre and that predicted for defects having electric dipole oscillators along the $\langle 110 \rangle$ lattice direction.

This evidence does not differentiate between different F_2-centre models[15,21]; other information will do so.

The important parameters describing the F_2-centre optical properties in four crystals are given in Table 4.4. It can be seen that the F_2-centre luminescence is qualitatively similar to the F_A-centre luminescence, since whether the excitation is into the $F_2(F)$-band or the F_2-band, only one emission band is observed. This emission band is polarized parallel to the F_2-centre axis and has a relatively small Stokes shift with respect to the F_2-band in absorption. Thus the emission band is due to transitions between the two states involved

Table 4.4 The F_2-band optical parameters in some alkali halides (After FOWLER[24] and HUGHES, POOLEY, RAHMAN and RUNCIMAN, 1967, A.E.R.E. Report R5604).

	Peak energy (eV)		Half-width (eV)	
Crystal	Absorption	Emission	Absorption	Emission
LiF	2·82	1·85	0·24	—
LiCl	1·91	—	0·19	—
NaF	2·47	1·88	0·12	0·25
NaCl	1·74	1·16	0·12	0·17
NaBr	1·61	—	0·17	—
KF	1·94	—	0·08	—
KCl	1·53	1·17	0·07	0·19
KBr	1·39	—	0·07	—
KI	1·23	—	0·07	—
RbF	1·72	—	0·09	—
RbCl	1·41	—	0·09	—
RbBr	1·29	—	0·06	—
RbI	1·14	—	0·07	—

in the F_2-band absorption. The polarization of the emission confirms that there is a high probability of radiationless transitions between the excited state reached by absorption in the $F_2(F)$-band and that state reached in the F_2-band. Thus the mechanism by which only a single emission band is observed for the F_2-centre is very different from the emission mechanism in the F_A-centre. The observations are consistent with the relaxed and unrelaxed excited states being very similar in the F_2-centre (hence the small Stokes shift), contrary to the situation in the F and F_A-centres where a large expansion of the wave function occurs in the relaxed excited state.[27]

Direct evidence for the divacancy model of the F_2-centre is obtained from experiments on the chemical equilibrium between F- and F_2-centres in crystals. Van Doorn investigated the relative strength of the F- and F_2-bands in KCl crystals additively coloured at 69 K in different pressures of potassium vapour. Faraday et al.[31] studied the F/F_2 equilibrium during irradiation at temperatures in the range 4–300 K. In both cases the F_2-centre concentration varied linearly with the square of the F-centre concentration. This result implies the reaction $2F \rightarrow F_2$ to be responsible for F_2-centre formation.

The final details of the atomic and electronic structure of the F_2-centre were confirmed by magnetic techniques. A number of measurements of magnetic susceptibility confirmed that the F_2-centre ground state is diamagnetic, the two electron spins being aligned antiparallel. Consequently electron spin resonance and ENDOR yield no information on the ground state. However, the conclusive proof of the F_2-centre model came from the electron spin resonance and ENDOR studies in an *excited* triplet state of the centre. The substantial population of the triplet state in KCl is achieved by intense radiation into the F-band (or low energy tail of the F-band) at about 90 K. The mechanism by which the triplet state becomes populated is not known. However the lifetime of this state is about 50 seconds and populations of the triplet level sufficient for magnetic resonance studies are readily achieved. Furthermore excitation of the triplet state is accompanied by suppression of the F_2-band and the appearance of a new band several tenths of an electron volt above the F_2-band. Consequently in the F_2-centre the triplet states are a little higher in energy than the corresponding singlet state. When a substantial concentration of F_2-centres are in the triplet state, new lines appear in the electron spin resonance spectrum, which decay with the same time constant as the changes in the optical spectrum. These new lines are all inhomogeneously broadened by an extensive hyperfine structure similar to that discussed earlier for the F-centres. In addition the spectrum shows a fine structure due to the defect having a total spin quantum number $S > \frac{1}{2}$. The angular dependence of the lines confirm the symmetry of the defect, since the line positions vary with orientation according as

$$H = H_0 + (M_s - \tfrac{1}{2})[D(3\cos^2\theta - 1) - 3E(\cos^2\theta - 1)\cos 2\gamma] \quad (4.13)$$

where θ is the angle between the defect axis and H_0, γ is the azimuthal angle, $H_0 = g\beta/h\nu$, D and E are the fine structure constants of the defect in units of magnetic field. In KCl the line positions are consistent with $S = 1$, $g = 1{\cdot}998$, $D = -161G$ and $E = +54G$. The value of S and the orientation dependence of the spectrum clearly define the number of electrons and the centre symmetry required by the F_2 centre model. In addition the value of D may be computed assuming that the fine structure originates in the mutual magnetic dipole interaction between the aligned magnetic spins, assuming that the two F-centres are situated at the nearest neighbour anion sites. The agreement between the experimental and theoretical values if regarded as good. The ENDOR spectrum reveals the hyperfine interaction with near neighbour cations and confirms the detailed atomic structure of the defect.

Recent experiments show that during certain treatments at low temperature the F_2-centres may be ionized to form F_2^+-centres or trap a third electron to form F_2'-centres. In general these centres have rather narrow absorption bands at lower energies than the F_2-bands. The electron-phonon coupling at these centres seems to be weaker than in the F_2-centre since at low temperatures zero-phonon lines and phonon-assisted sidebands are associated with their absorption and emission bands. Unlike the F'-centre, which has excited states degenerate with the conduction band, the F_2'-centres must have bound excited states otherwise the sharp lines would not be observed. This is quite reasonable since the extra electron is more strongly bound to the aggregate centre than in the F'-centre since it is shared among a greater number of vacancies.

Other aggregate centres

F_3- and F_4-centres represent a further stage in the F-centre aggregation process. The structure of the F_3-centres is well understood; the situation is less satisfactory for the F_4-centres. Both the $F_3(1)$-band and $F_3(2)$-band shown in Fig. 4.12 are associated with different electronic transitions within the same F_3-centre.[23] The electronic states of the F_3-centre are strongly influenced by the symmetry of the crystal field which they experience. Consequently the only allowed electronic states of the F_3-centre are orbital singlet and orbital doublet states. The ground state of the F_3-centre has a resultant spin $S = \frac{1}{2}$ and is an orbital doublet, a property which gives rise to interesting complexities in both the optical spectrum and electron spin resonance spectrum of this defect.[32,33]

At low temperatures the $F_3(2)$-bands show a well developed zero-phonon line and ancilliary structure. This structure has been used by numerous authors as a sensitive probe to uncover the nature of the defect structure. Since the zero-phonon line is narrow, many experiments become possible which are not generally feasible for broad band studies. Especially prominent in this respect is the application of an external perturbation, e.g. a uniaxial stress, which may remove certain types of degeneracy associated with the defect. In the simplest case the defect site has cubic symmetry and the orbital degeneracies of the electronic states may be removed by the applied perturbation. Some defects, however, have only orbital singlet states and then only *orientational degeneracy* is removed. This corresponds to the defect being anisotropic with a number of equivalent orientations of the defect axis. When a stress is applied some orientations which are degenerate in the absence of stress become inequivalent, and the zero-phonon lines split into a number of components. Analysis of the splittings for a number of stress configurations relative to the defect axis shows that the F_3-centre has both orientational degeneracy due to its trigonal symmetry, and orbital degeneracy associated with transitions from a doubly degenerate ground state to an orbital singlet excited state.[32–35] This observation substantially confirms the model of the defect shown in Fig. 4.13. The most conclusive evidence for this model, however, is obtained from electron spin resonance. In the ground

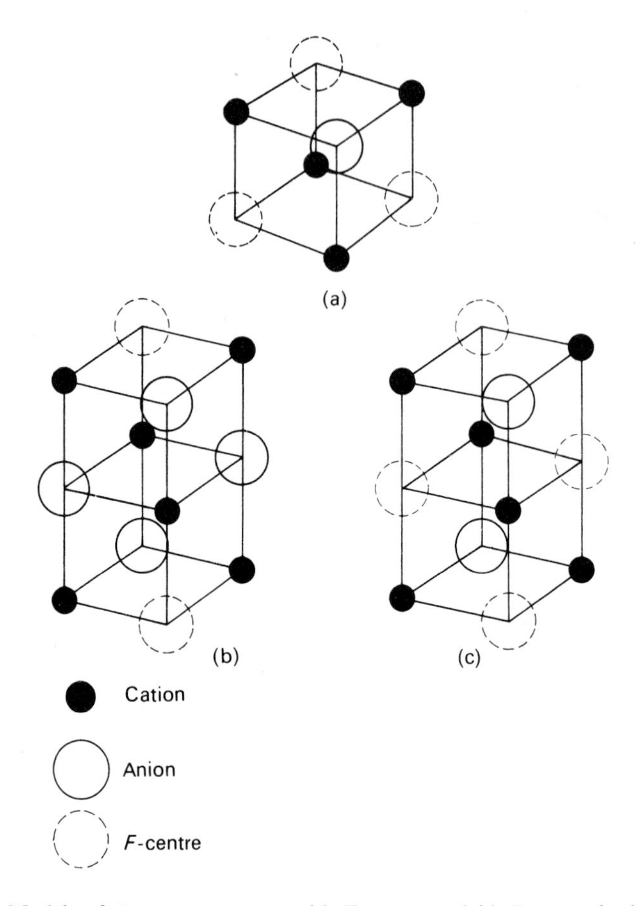

Fig. 4.13 Models of F-aggregate centres (a) F_3-centre and (c) F_4-centre both after PICK (1960. *Zeit f. Phys.* **159,** 69) and (b) F_2-centre proposed by JOHANSON, LANZL, VON DER OSTEN and WAIDELICH. 1965. *Phys. Rev.* **15,** 110.

state the F_3-centre has total spin momentum with $S = \frac{1}{2}$, and an electron spin resonance signal is expected. Unfortunately the two-fold orbital degeneracy of the ground state leads to complications which result in the spectrum being broadened beyond detection under normal conditions. A very careful study by Krupka and Silsbee[33] revealed that the spectrum may be observed in KCl below 4·2 K when the crystal is subjected to a large external stress. The spectrum is characterized by $g_{\perp} = 2{\cdot}06$ and $g_{\parallel} = 2{\cdot}00$; these values agree well with theoretical predictions based upon the F_3-model of the F_2-centre. Further confirmation in KCl was reported by Seidel *et al.*[36] who excited these centres into higher energy levels with maximum total spin $S = 3/2$. These spin quartet states are metastable with a lifetime of 14·5 seconds. Both electron spin resonance and ENDOR studies on these metastable quartet states define the detailed electronic and atomic structure of the F_3-centre without ambiguity.

The F_4-centres have received less attention and their atomic and electronic structure are still problematic. However, their symmetry properties have been determined from uniaxial stress experiments on the optical zero-phonon lines. The zero-phonon line attendant upon the F_4-band in NaCl affords an especially clear example of the removal of orientational degeneracy in optical spectroscopy. Figure 4.14 shows the splitting of this line for stress along the [110] axis and for two different viewing directions. The number and intensities of the components are in good agreement with expectations for a centre with monoclinic symmetry and electric dipole moments along a ⟨112⟩

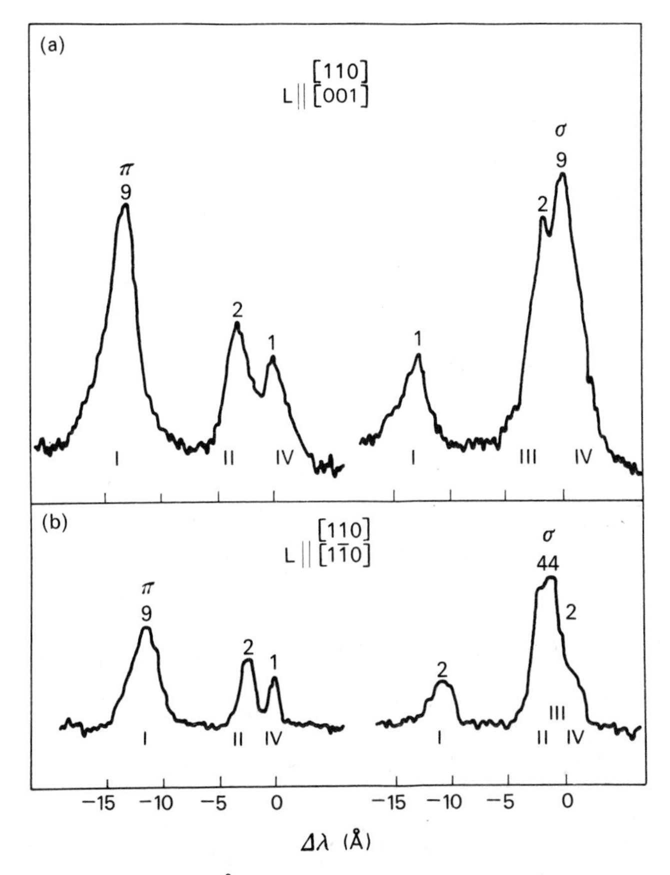

Fig. 4.14 Splitting of the 8373 Å F_4-line in NaCl under a [110] stress and two different viewing directions. Predicted intensities for monoclinic symmetry appropriate to 4.13(b) and (c) are indicated. (After HUGHES. 1966. *Proc. Phys. Soc.* **87**, 535.)

direction. Although we know the symmetry of the F_4-centre it is not possible to differentiate between the two models (b) and (c) in Fig. 4.13.

The phonon assisted sidebands in the F_4-centre absorption spectrum in NaCl are especially well resolved. These sidebands should in general reflect the critical points in the phonon spectrum of the pure lattice. A comparison between the sideband spectrum of the F_4-centre and the theoretical density of states curve in Fig. 4.15 reveals a reasonable correspondence between the

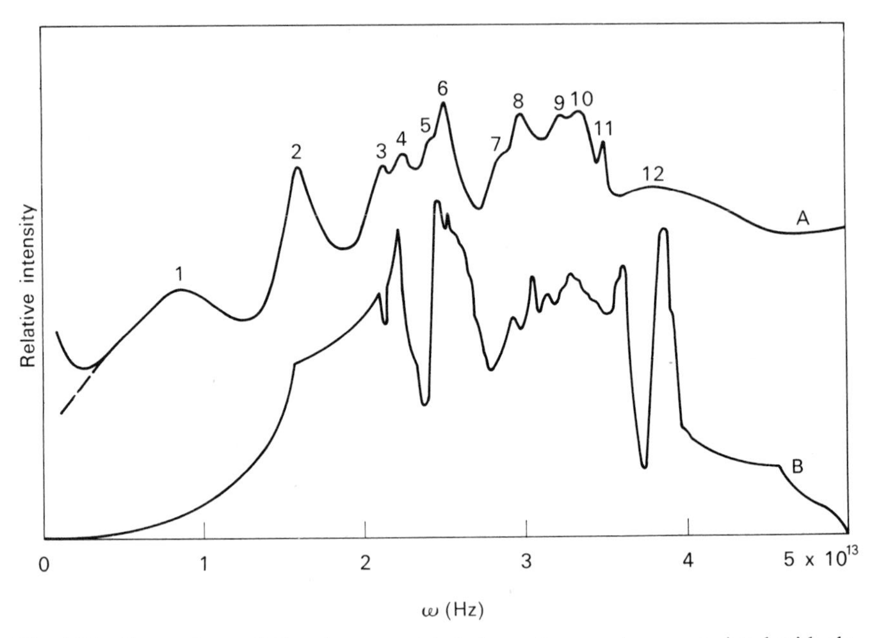

Fig. 4.15 Comparison of the phonon-assisted absorption spectrum associated with the 8373 Å F_4-line in NaCl (A) with the theoretical vibrational density of states (B). (After HUGHES. 1966. *Proc. Phys. Soc.* **88,** 449.)

peaks in the two curves. Hughes[37] interprets the vibronic spectrum as due to one phonon interactions with a number of normal lattice modes superimposed on a strong interaction with a localized mode (peak 1 in Fig. 4.15).

4.7 Trapped hole centres in the alkali halides

Despite playing an important role in the development of colour centre research, trapped hole centres are not as well understood as the electron excess centres discussed above. V-centres are produced in the alkali halides by additive colouration with excess halogen or by irradiation. In general the V-bands labeled V_1, V_2, V_3, etc. lie to shorter wavelengths than the F-bands; they are usually quite broad compared with the F-band and F-aggregate bands. Irradiation at room temperature, or warming to room temperature subsequent to low temperature irradiation, produces V_2- and V_3-bands at higher photon energies than the V_1-band, at the expense of the V_1-band. Usually if an alkali halide crystal is irradiated at about 90 K, the V_1-band predominates, although X_2^--centres are also produced. In KCl the V_1- and X_2^--bands are almost co-incident at about 3·40 eV, and this led to an earlier incorrect assignment of the V_1-band to the Cl_2^--centre.[21] However, X_2^--centres have weak bands in the range 1·55–1·65 eV in addition

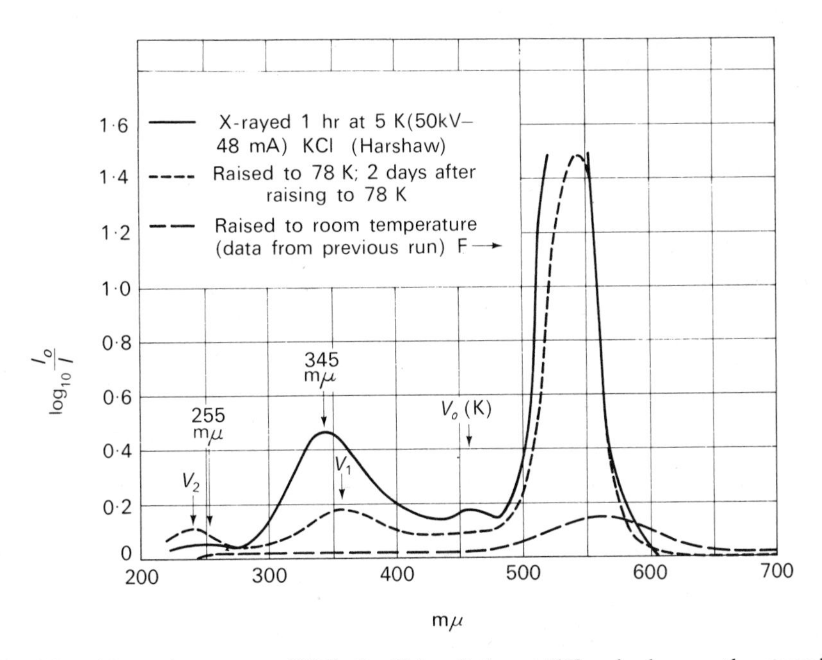

Fig. 4.16 Absorption spectra of KCl after X-irradiation at 5°K and subsequently warmed. The peak at 345 mμ is the *H*-band. (After Duerig and Markham. 1952. *Phys. Rev.* **88,** 1043.)

to strong bands in the near ultra-violet.[24] When X-irradiation and optical measurements are carried out at even lower temperatures (5 K) yet another band, the *H*-band, is created almost co-incident with the V_1-band. Warming to about 200 K bleaches the *H*-band and the V_1-band is enhanced. Figure 4.16 shows that a concomitant effect of this bleach is that the *F*-band also is bleached and the V_2-band also arises. The understanding of these processes came about mainly as a result of the electron spin resonance studies of Känzig and his co-workers.[38] Subsequent to this work the results of optical measurements are much more amenable to interpretation.

The first trapped hole centre recognized on account of its electron spin resonance spectrum was the X_2^--centre.* The essential ingredients of the spectrum are that the centre has $S = \frac{1}{2}$; there is a large g-tensor anisotropy and positive g-shift, and a strong hyperfine interaction with the two halide components of the defect. The data obtained from the ESR study permitted a very detailed analysis of the electronic structure of the defect. The centre was shown to consist of a halogen molecule ion (e.g. F_2^-), situated as shown in Fig. 4.2 at a single halogen ion site and aligned with the molecular axis parallel to the crystal [110] axis. A strong localization of the hole on the two halogen ions and the strong covalent bond required to hold the ions together confirms that the description of the defect in terms of molecular ions in the crystal is appropriate. Assuming the molecular nature of the centre, the g-shift, hyperfine structure and spectral position of the optical transitions may be computed with reasonable accuracy.[39] Having recognized the molecular symmetry of the defect, it is reasonable to expect dichroic optical absorption of the centre under certain conditions. Thus dichroism may be observed as a result of re-orientation of the X_2^--centres even at temperatures as low as 4 K. Experiments with polarized light then become especially useful.

The value of dichroism and polarization methods in determining the symmetry characteristics of lattice defects was further demonstrated by experiments on the H-band. The investigators measured the dichroism of the H-band after bleaching with polarized H-band light. The results showed that this centre also has an axis of symmetry parallel to the [110] crystal axis. The later ESR studies on H-centres in KCl and KBr indicated the crowdion configuration depicted in Fig. 4.2. Compared with the X_2^--centre, the trapped hole on the H-centre has an interaction with four halogen atoms distributed over three anion sites. Thus the H-centre differs from the X_2^--centre since an interstitial site is involved. The two centres are, however, both aligned in the $\langle 110 \rangle$ direction and their electron spin resonance spectra are very similar. This is shown in Fig. 4.17 for V_K- and H-centres in LiF, where the interacting nuclear isotopes are the 100% abundant ^{19}F with $I = \frac{1}{2}$. In Fig. 4.17 H_0 is directed along a cube axis: the strongest lines are due to the X_2^--centre, the remainder to the H-centre. The trapped hole in the X_2^--centre is distributed over two equivalent nuclei and the hyperfine interaction for two equivalent nuclei with $I = \frac{1}{2}$ gives the resultant spectrum. In the H-centre the unpaired electron spin is distributed over a further two nuclei giving rise to a second splitting reduced by a factor 10 compared with the major splitting. This result is consistent with the unpaired electron spin spending 4 to 10% of its time on these more remote ions. In this sense there is only a very weak binding between the F_2^--molecule ion and these two further ions which constitute the H-centre.

It is now possible to understand at least partially the bleaching behaviour, shown in Fig. 4.16. The H-centre structure is complementary to the F-centre, and if allowed to interact they will mutually annihilate one another. There is now strong evidence that the V_1-centre in KCl and KBr consists of

* The X_2^- centre was formerly referred to as the V_K centre.

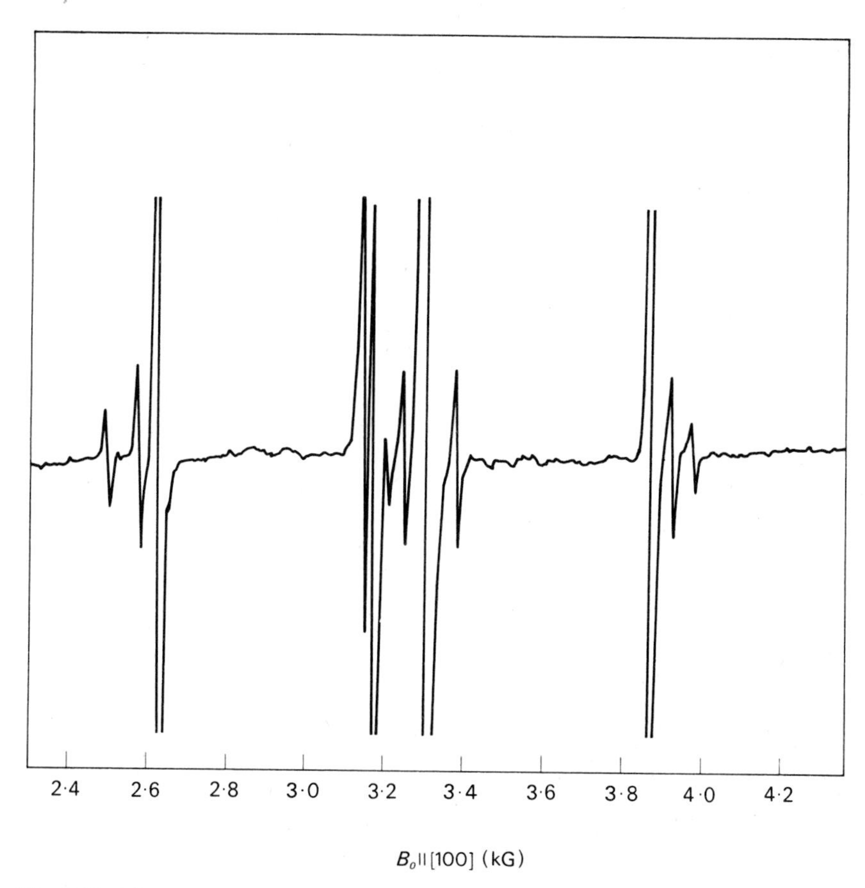

Fig. 4.17 The ESR spectrum at X-band of H-centres and X_2^--centres in LiF at 77 K with H_0 field along a ⟨100⟩ direction. All lines that go off chart belong to the X_2^--centres, the others to the H-centres. (After WOODRUFF and KANZIG. 1958. *J. Phys. Chem. Sol.* **5**, 268.)

an H-centre associated with an impurity Na^+ ion. Thus the decreased F-band and H-band and increased V_1-band are due partially to H-centre/F-centre recombination and partially to H-centre/Na^+ association. The structure of the centre giving rise to the V_2-band is not understood.

4.8 Insulators other than the alkali halides

Colour centre phenomena in the alkali halides are greatly simplified by the simple crystal structure and essentially ionic bonding. An extension of colour centre investigations to other materials is highly desirable in order that effects due to valency, bond-type and crystal structure may be critically assessed.

Systematic studies of colour centres in the caesium halides have been made: F-centres and their aggregates are qualitatively similar to the alkali halides despite their different crystal structure. The silver halides which are iso-structural with the alkali halides have been investigated mainly from the viewpoint of image-forming process in photographic emulsions.[40] The alkaline earth oxides are another important class of materials, since they are the divalent structural analogies of the alkali halides. In these compounds, oxygen ion vacancies possess extra potential for trapping electrons, and bound excited states exist for a defect in which two electrons are trapped at an anion vacancy—the F^+-centre. The paramagnetic F^+-centres have been extensively investigated using electron spin resonance. Such studies demonstrate that the electron is much more strongly concentrated in the anion vacancy than in the alkali halides. Nevertheless in MgO resonances due to hyperfine interaction with shells more remote than the first shell of cations have been observed using ENDOR. In MgO extensive studies of optical zero-phonon lines observed in neutron irradiated crystals have resulted in suggestions that vacancy aggregate centres analogous with the F_2- and F_3-centres in the alkali halides may be present. For the same structural arrangement of three vacancies the F_3^+, F_3, and F_3^--centres may contain 5, 6 and 7 electrons in MgO. There are also a number of V-centres produced by X-irradiation in the alkaline earth oxides, although their presence appears to be dependent upon the impurity content of the crystal. It is clear, however, that the divalent host lattice introduces important differences between defect states in the oxides and in the alkali halides.

In the alkaline earth fluorides the F-centre differs slightly from the alkali halide F-centre in respect of the symmetry around the anion site. The effects are easily recognized in the ESR and ENDOR spectra of F-centres in these structures. The structure of the F_2^--centre has also been determined in the alkaline earth fluorides. Finally we comment that a large number of measurements have been reported on colour centres in less symmetrical lattices, including the alkali nitrides, the ammonium halides, aluminium oxide, quartz, mixed oxides (e.g. $CaWO_4$) and fluorapatite $Ca_{10}(PO_4)_6F_2$. Some effects of covalency in colour centre properties are discussed in the following chapter on defects in semiconductors.

Appendix A4

The Sonder–Sibley notation rules for point defects in insulators

A consistent method of naming defects applicable to all polar crystals has been proposed by Sonder and Sibley, 1972, *Point Defects in Solids* (Ed. Crawford and Slifkin, Plenum, New York). The scheme proposed is a compromise between retaining the notation used in the literature mainly for alkali halides and adopting one which is somewhat less restrictive. Overleaf are reproduced the general rules which can be used for naming new defect centres in a consistent fashion whenever their configuration becomes known.

RULES

1a The name F centre is reserved for negative-ion vacancies containing the *same number of electrons as the charge of the normal lattice ion.* Thus, in singly charged polar crystals (e.g., alkali halides) the F centre is a vacancy with one electron and in doubly charged polar crystals, as for example MgO, the F centre is a vacancy with two electrons.

1b The name V centre is reserved for positive-ion vacancies whose neighbours contain the same number of holes as the charge of the normal lattice ion that is missing.

1c The name H centre is reserved for a negative-ion interstitial atom that has combined with a lattice ion so that a molecular ion shares a normal lattice site. The H centre has no net charge as compared with the perfect lattice.

2 If an F, V, or H centre is adjacent to an impurity, the composite defect can be specified by subscripts following the F, V, H notation. Thus, an F_A (or F_{Na}) is an F centre adjacent to an impurity cation (or particularly Na); V_{OH} is a positive-ion vacancy that is next to a substitutional OH^- impurity.

3 An aggregate of single point defects is specified by a number subscript. Thus, for example, three adjacent F centres are labeled F_3.

4 The defects specified in 1–3 all have the same charge as does the perfect lattice. If the charge is different from that of the perfect lattice, a superscript follows the defect symbol. For example, a negative-ion vacancy in an alkali halide that does *not* contain an electron would be labeled F^+, since the empty vacancy is one electronic charge more positive than a negative ion that would be there in the perfect halide lattice.

5a Centres for which the number of ions (atoms) is equal to the number of lattice sites are not ionic defects in the normally accepted sense and are not given letter designation (as, for example, V centre). They are designated, as has been done in the recent literature, by specifying the ions or molecules involved. Thus, a self-trapped hole in KCl or NaCl is a $[Cl_2^-]$ centre and a hole trapped at a lattice Cl^- and a neighbouring substitutional Br^- is a $[ClBr^-]$ centre. This notation can also be used for substitutional impurities; for example, for iron in MgO the notation would be $[Fe^{2+}]$ or $[Fe^{3+}]$.

5b Brackets are used around all centres that are written in terms of their chemical species. This is necessary to prevent confusion with defect centres like F or H centres, which could be confused with fluorine or hydrogen impurities.

5c The superscript giving the valence of the centre should be placed inside the bracket since it is descriptive of the centre and does not reflect the charge, with respect to the crystal, of the defect. For example, a $[ClBr^-]$ has a net positive charge in an alkali halide as does an $[Fe^{3+}]$ in a divalent oxide. If desirable, the charge of the defect with respect to the lattice can be placed outside the bracket (for example, $[ClBr^-]^+$).

5

Defects in Crystalline Semiconductors

5.1 Some characteristics of semiconductors

Semiconductors have electrical resistivities in the range 10^{-2} to 10^{7} ohm cm, which are intermediate between those of typical metals and typical insulators. Prior to the last decade the experimental and theoretical emphasis on silicon and germanium led to their exploitation as solid state rectifiers. Recently intermetallic compounds have assumed importance. Their uses vary widely: compounds such as indium antimonide and indium arsenide have useful electrical properties, lead sulphide and lead selenide are important photoconductors, while numerous Group II–VI compounds are valuable for their optical properties.

Bonding in elemental semiconductors

The free atoms of the Group IV elements carbon, silicon, germanium and tin have four valence electrons in the electronic configuration $(ns)^2(np)^2$ outside the stable ion core. In crystalline solids these $(ns)(np)$ orbitals overlap resulting in four electrons/atom being contributed to an integral number of energy bands, the so-called valence bands. The band next highest in energy, the conduction band, is unoccupied by electrons at 0 K. It is separated from the valence bands by an energy gap, the magnitude of which is crudely correlated with the bond strength. The bond strength is more accurately assessed in terms of the amount of overlap between the atomic orbitals on neighbouring atoms. The strongest bonds are formed when overlap of the atomic orbitals is optimized by the formation of (sp^3) hybridized bonding orbitals. Detailed quantum mechanical consideration of these hybrid bonds shows that appropriate linear combinations of the atomic (ns) and (np) wave functions yield four equivalent bonding orbitals concentrated along directions from the atom to the corners of a regular tetrahedron (Fig. 5.1). Each orbital contains one electron per atom and thus the four valence electrons in the free atom configuration become equivalent in the solid as far as bonding is concerned. The overlap between orbitals on the central atom and the atoms at the corners of the tetrahedron leads to covalent bonding. At least one atom in every unit cell must complete an octet of *s*-*p*

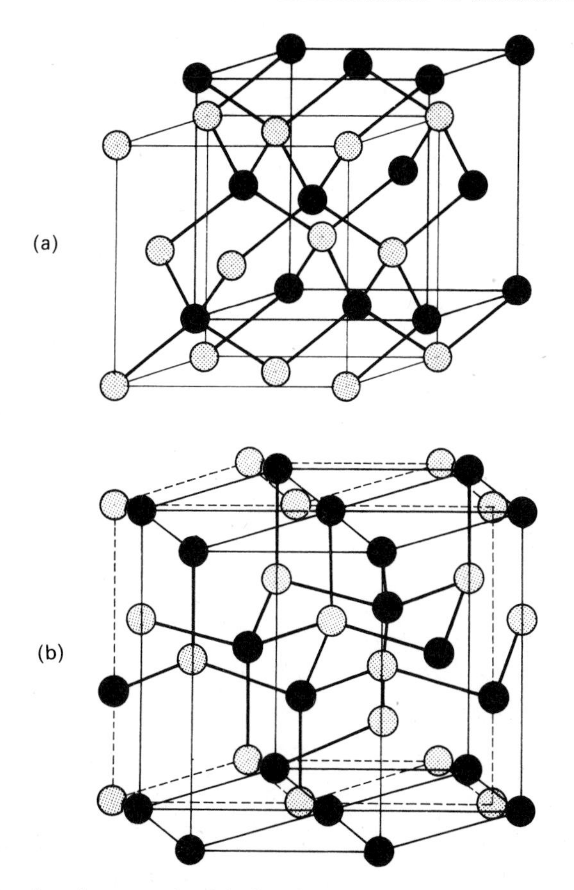

Fig. 5.1 Showing the network of (sp^3) hybrid bonds which permeate (a) both the face-centred cubic diamond structure (all atoms identical) and the zinc blende structure, and (b) the hexagonal wurtzite structure.

electrons by forming covalent bonds, each of which involves two electrons with mutually opposed spins, one electron being contributed by each atom. Thus there is a fairly high electron density between the ion cores. In Fig. 5.1 these (sp^3) hybrid bonds are shown to form a continuous network throughout the crystal structure.

It should be emphasized that the energy gap between valence and conduction bands is not constant at all points in the Brillouin zone. Consider as an example Fig. 5.2 which shows the theoretical band structure of germanium. The top of the valence band (Γ'_{25}) at the point $k = 0$ in the Brillouin zone is separated from the lowest point (L_1) in the conduction bands by 0·66 eV: in Ge there are four equivalent conduction band minima situated at zone boundaries in the $\langle 111 \rangle$ directions. Similar minima occur in Si about

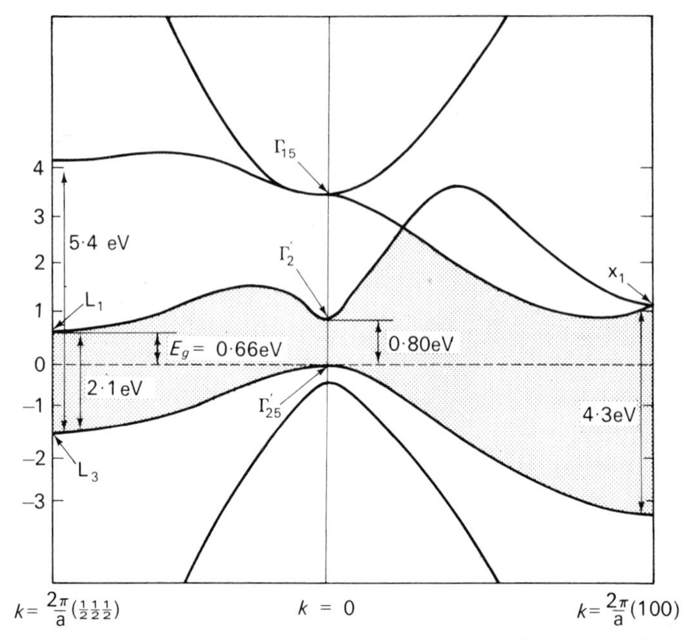

Fig. 5.2 The calculated band structure of germanium near the band gap, adjusted slightly to fit the known experimental splittings. (After KITTELL. 1966. *Introduction to Solid State Physics*. 3rd Edition. Wiley, New York.)

85% of the way between $k = 0$ and the zone boundary along the zone $\langle 100 \rangle$ direction. The band gaps in Table 5.1 represent the indirect gaps between the maxima in the valence band and the minima in the conduction bands.

Conductivity in semiconductors

In semiconductors at 0 K the conduction bands are empty and such solids behave like insulators. Conductivity occurs when the temperature is sufficiently elevated that covalent bonds are broken by thermal agitation, producing electrons and holes in equal numbers. Referring to Table 5.1 it is evident that diamond and grey tin are of no practical use as semiconductors since the covalent bonds in diamond are too strong and in grey tin too weak. Consequently diamond remains an insulator even at quite high temperatures, and grey tin is a conductor at very low temperatures.

In perfectly pure crystals, the thermal excitation of electrons across the band gap (E_g) leaves positive holes in the valence band, which also act as current carriers. The total conductivity, σ, of this *intrinsic semiconductor* is

$$\boxed{\sigma = e(n_e\mu_e + n_h\mu_h)} \tag{5.1}$$

Table 5.1 Some Properties of Semiconductors

Material	Structure	Lattice constant(s) Å	Melting point °C	Band gap eV	Mobilities $cm^2/V/sec$ Electron μ_e	Hole μ_h	Ionization energy E_i (eV)
Diamond	F.C.C. diamond	3·56	3 500	5·33	1 800	1 200	—
Si	F.C.C. diamond	5·40	1 420	1·14	1 600	400	0·039 (Sb), 0·065 (Ga)
Ge	F.C.C. diamond	5·65	936	0·67	3 800	1 800	0·01 (Sb), 0·011 (Ga)
Sn (Grey)	F.C.C. diamond	6·48	232	0·08	2 500	2 400	—
AlP	Cubic Zinc blende	5·45	2 000	2·5	—	—	—
AlAs	Cubic Zinc blende	5·62	1 600	2·3	—	—	—
AlSb	Cubic Zinc blende	6·1356	1 065	1·72	200–450	200–450	0·07 (Te)
GaP	Cubic Zinc blende	5·4506	1 467	2·25	—	20	—
GaAs	Cubic Zinc blende	5·6535	1 238	1·4	600	400	0·05 (Zn)
GaSb	Cubic Zinc blende	6·0955	706	0·78	3 000	650	0·02 (Zn)
InP	Cubic Zinc blende	5·8688	1 058	1·25	3 400	650	—
InAs	Cubic Zinc blende	6·0585	942	0·33	23 000	100	—
InSb	Cubic Zinc blende	6·4789	530	0·23	77 000	1 250	0·008 (Zn)
ZnS‡	Cubic Zinc blende	5·423	1 850*	3·6	—	—	—
ZnSe	Cubic Zinc blende	5·65	1 500	2·6	—	—	—
ZnTe	Cubic Zinc blende	6·07	1 240	0·8	—	—	—
CdS‡	Cubic Zinc blende	5·82	1 750	2·42	210	—	0·03 (Ga)
CdSe	hex. ZnO (wurtzite)	$a = 4{\cdot}3$, $c = 7{\cdot}02$	1 350	1·74	100	—	—
CdTe	cub. ZnS	6·41	1 040	1·45	950	90	—
PbS	cub. NaCl	5·97	1 114	0·35	600	1 000	0 02 (Cu)
PbSe	cub. NaCl	6·14	1 065	0·27	900	1 000	—
PbTe	cub. NaCl	6·34	920	0·30	17 000	1 000	—
ZnO	hex. ZnO (wurtzite)	$c = 5{\cdot}1948$, $a = 3{\cdot}2427$	1 800	3·2	175	—	—
CdO	cub. NaCl	4·689	900†	2·3	20	—	—
Cu_2O	cubic	4·26	1 235	2·1	—	—	—
TiO_2	tetr. rutile	$a = 4{\cdot}58$, $c = 2{\cdot}95$	2 000	3	—	—	—

* Melting point measured under high pressure of S vapour (100 atm) † Decomposes at this temperature
‡ Also crystallizes in wurtzite crystal structure.

where n and μ refer to the concentrations and mobilities respectively of electrons (e) and holes (h). In general $\mu_e > \mu_h$ (see Table 5.1). Since in intrinsic semiconductors $n = n_e = n_h$ Equ. 5.1 becomes

$$\sigma = ne(\mu_e + \mu_h)$$

The concentration of free carriers n is a function of both E_g and T, and the temperature dependent form of Equ. 5.1 may be written as

$$\boxed{\sigma = AT^B \exp\left(-\frac{E_g}{2kT}\right)} \tag{5.2}$$

A knowledge of the constants A and B is contingent upon establishing a suitable microscopic model for electron scattering. The dominant term is almost always the exponential term and E_g is deduced from the slope of the curve obtained by plotting $\log_e \sigma$ as a function of T^{-1}. Equation 5.2 shows that in contrast to metals, semiconductors are characterized by a negative temperature coefficient of resistivity.

Impurity effects

The low temperature conductivity of many semiconductors indicates that they contain impurities which are dissolved either substitutionally or interstitially in the lattice. The impurities may or may not be electrically active. Semiconductors which contain electrically active impurities may belong to one of two classes of *extrinsic semiconductor*. For example, Si and Ge containing Group III impurities are *p-type semiconductors* since they contain an excess of positive charge carriers at low temperature. Group III elements introduce *acceptor levels* just above the valence bands and so provide energy levels into which electrons may be thermally excited from the valence band. The positive holes left behind in the valence band then predominate as current carriers. A number of Group I and transition metal elements also result in p-type characteristics. Paramagnetic resonance studies have shown that the valence states of transition metal impurities vary according to the position of the Fermi level.

Group V elements in Si and Ge contribute one electron per atom additional to those required for hybrid bond formation between nearest neighbour atoms. These elements provide *donor levels* within the band gap, very close to the conduction band minimum. Consequently these extra and weakly bound electrons on the donor atoms require the expenditure of little thermal energy for their promotion into the conduction band. Thus conductivity is via the transport of negative charge carriers in *n-type semiconductors*. Wannier[41] has treated shallow donors using a model in which the extra electron is bound to a spherically symmetric uni-positive charge, embedded in a solid dielectric material. Thus the model represents the impurity as a quasi-hydrogenic atom in a continuous dielectric material. The solutions of the appropriate Schrödinger equation are obtained by scaling the solutions for the hydrogen atom according to the dielectric constant, ϵ. Thus the binding energy of the electron to the atom, E_b, is:—

$$E_b = \frac{-e^4 m^*}{2n^2\hbar^2\epsilon^2} = \frac{-13{\cdot}6}{n^2\epsilon^2}\left(\frac{m^*}{m}\right) \text{ eV} \tag{5.3}$$

and the orbital radius of the electron

$$r = \frac{n^2\hbar^2\epsilon}{e^2 m}\left(\frac{m}{m^*}\right) = \frac{0{\cdot}53 n^2\epsilon}{(m^*/m)} \text{ Å} \tag{5.4}$$

where m^*/m is the ratio of the effective mass to rest mass of the electron. Assuming an experimental value for $m^*/m = 0{\cdot}25$[42] and $\epsilon = 11{\cdot}7$ we obtain $E_b = 0{\cdot}025$ eV and $r = 70$ Å. Table 5.1 shows that the result for the binding

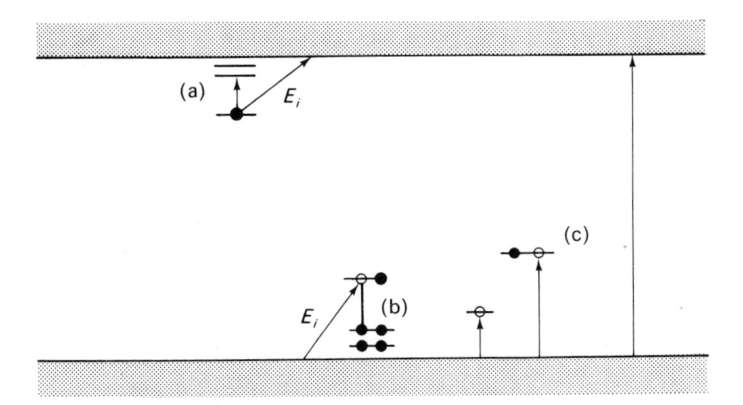

Fig. 5.3 Schematic representation of energy levels of donors and acceptors in semi-conductors: (a) simple donor, (b) simple acceptor (c) double acceptor.

energy is in reasonable agreement with experimental values of ionization energies. The magnitude of the orbital radius implies that the wave function of the electron on the Group V atom extends over a large number of silicon sites.

In the schematic energy level diagram in Fig. 5.3, E_i refers to either the energy required to promote electrons from the valence band into acceptor levels or from donor levels into the conduction band. Since $E_i \ll E_g$, extrinsic conductivity is important at much lower temperatures than intrinsic conductivity. The interplay between impurity-activated conductivity and intrinsic conductivity may result in there being three regions in the conductivity temperature curves. The processes which produce this behaviour are shown schematically in Fig. 5.4. At low temperatures the number of thermally activated extrinsic sources of conductivity is determined by the Boltzmann equation,

$$n_e \quad \text{or} \quad n_h = c \exp\left(-\frac{E_i}{2kT}\right)$$

and consequently the resistivity, ρ, decreases with increasing temperature. This process is rapidly exhausted with increasing temperature when all the

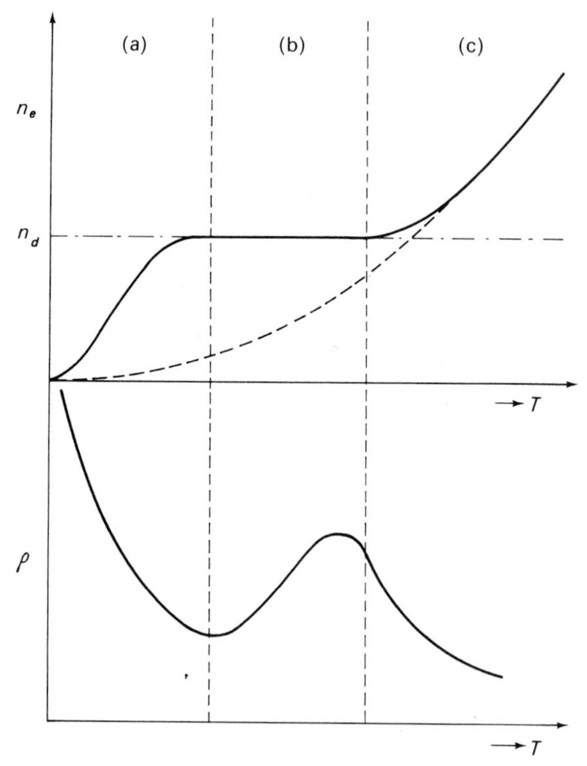

Fig. 5.4 Schematic variation of the density of conduction electrons n_e and resistivity ρ of an extrinsic semiconductor as a function of temperature n_d is the concentration of donor centres. Regions (a) and (b) involve extrinsic behaviour while region (c) gives the extrinsic behaviour at high temperature.

impurity sources have been activated. For materials with a large band gap there may then be little change in the carrier concentration with further increase in temperature. Consequently over some temperature range the resistivity may actually increase due to the reduction of the free carrier mobility with increasing temperature. At still higher temperatures intrinsic conductivity dominates and the resistivity decreases again.

Compound semiconductors

The continuing search for new semiconductors has stimulated much investigation of the compounds formed between elements in Groups III and V of the Periodic Table (e.g. GaAs, InSb) and in Groups II and VI (e.g. ZnO, CdS). These materials have the wurtzite or zinc blende structure both of which are shown in Fig. 5.1 and are related to the diamond structure. Although these compounds are formed between elements in different columns of the Periodic Table, covalent bonding predominates since the average number of bonding electrons per atom is the same as for Si and Ge. The ionic contribution to bonding increases both the melting point and the band gap relative to the elements. Examination of the properties of the elemental and compound semiconductors, given in Table 5.1, makes this particularly evident. Since ionicity decreases and the amount of metallic character increases* with increasing atomic number so the band gap diminishes. These compounds have band structures which are similar to the elemental semiconductors, differing mainly in the positions of the conduction band minima. An ideally simple semiconductor band structure would have at $k = 0$ in the Brillouin zone, a single maximum in the valence band and a single minimum in the conduction band. No semiconductor has such a simple band structure, although cadmium sulphide closely approaches this *idealized* situation.

The defect structure and impurity effects in compound semiconductors differ from those in silicon and germanium. Non-stoichiometry becomes important especially in the II–VI compounds. Furthermore donor and acceptor levels may be introduced on both sublattices. Zinc and cadmium are the most important acceptors in the III–V compounds and both substitute on the trivalent sublattice. The Group VI elements sulphur, selenium and tellurium act as donors on the pentavalent sublattice. Impurities from Group IV behave as acceptors or donors, substituting on one of the two sublattices preferentially. Generally impurities occupy sites on the sublattice which offers most space.

A final class of material not considered here is also being actively investigated. We refer to the large number of organo-metallic compounds which have semiconducting properties.

5.2 Diffusion and point defects in thermal equilibrium

Silicon and germanium

The relationship between diffusion and the point defect concentration at equilibrium is especially important in semiconductor technology since

* This effect can be seen in the elements in any one of the columns IVB, VB or VIB of the Periodic Table.

diffusion methods are used extensively in producing *p-n* junctions. Considerable research has made it possible to control accurately both the depth and sharpness of *p-n* junctions by diffusion methods. General-purpose *p-n* junctions usually consist of a thin layer of heavily doped Si or Ge (low resistivity) on a matrix of pure Si or Ge (high resistivity). However, high frequency transistors require a thin layer of high resistivity material on a low resistivity matrix and are made differently. Usually the pure Si or Ge is deposited epitaxially on a matrix of heavily doped Si or Ge, by thermal decomposition of a gaseous compound. The temperature is kept low to preclude any diffusion from the low resistivity matrix into the epitaxial layer. The 'thermal conversion' of *n*-type Ge to *p*-type after quenching from high temperature is yet another example of the importance of diffusion. This effect is concerned with substitutional Cu impurities forming a *p-n* junction which diffuses from the surface into the interior of the crystal.

The defect concentrations at thermal equilibrium in pure Si and Ge are much smaller than those in pure metals with similar melting points. Direct detection of the intrinsic point defects using the techniques outlined in Chapter 2 is not possible. Instead the properties of intrinsic point defects are often inferred from diffusion measurements. Since the defect concentrations are very low it is found that self-diffusion occurs only slowly and account should be taken of the possible effects of impurities upon the self-diffusion rate of the matrix. Self-diffusion in Si and Ge is slow mainly because the activation energies for self-diffusion are so large (see Table 5.2). These large activation energies have important implications in respect of the diffusion mechanisms. Extrapolation of the self-diffusion data given in Table 5.2 to the melting temperature using Equation 1.6 gives diffusion coefficients D (m.t.) which are only of order 10^{-12} cm^2 s^{-1}, compared with values of 10^{-8} to 10^{-9} for face-centred cubic metals. Consequently Seeger and Swanson[43] conclude that although defect-induced diffusion undoubtedly occurs in valence crystals, the monovacancy mechanism which persists in many metals is of less importance.

Lower limits of the defect concentrations may be estimated in the following way. The self-diffusivity D^d induced by migration of intrinsic lattice defects is written in terms of the molar defect concentration η and a pre-exponential factor D_0^d as,

$$\boxed{D^d = D_0^d \eta} \tag{5.5}$$

where $D_0^d \leqslant a^2 \nu_{\text{eff}}$ according to Equ. 1.5. Since the effective jump frequency ν_{eff} is unlikely to be greater than the Debye frequency $\nu_D = k\theta_D/\hbar$, D_0^d is given by the inequality,

$$D_0^d < a^2 k\theta_D/\hbar$$

hence

$$\boxed{D^d < \frac{a^2 k\theta_D}{\hbar}\eta} \tag{5.6}$$

Table 5.2 Diffusion coefficients $D = D_0 \exp - (E_D/kT)$ in semiconductors*

Material	Type and/or dopant	Temperature range °C	D_0 cm^2/s^{-1}	E_D eV	D(m.p.) cm^2/s^{-1}
Si	Intrinsic	1220–1400	1800	4·86	3×10^{-12}
	B-acceptor		3–10	3·6	—
	Ga-acceptor		40	3·9	—
	P-donor		29	3·88	—
	As-donor		70–150	4·3	—
	Li		$(2{\cdot}2–4{\cdot}4) \times 10^{-3}$	0·7	—
	Fe		$6{\cdot}2 \times 10^{-3}$	0·86	—
Ge	Intrinsic	750–870	10·8	3·0	6×10^{-12}
	n-type (6×10^{18})	750–870	0·13	2·6	—
	p-type (5×10^{19})	750–870	25 000	3·8	—
	B-acceptor		$1{\cdot}6 \times 10^9$	4·6	—
	Ga-acceptor		40	3·14	—
	P-donor		2·5	2·48	—
	As-donor		3	2·4	—
	Li		$(1–9) \times 10^{-3}$	0·5	—
	Fe		0·13	1·08	—
GaAs	Intrinsic	1125–1250	10^7 (Ga)	5·6	8×10^{-12}
			4×10^{21} (As)	10·2	6×10^{-14}
	Cd-acceptor		0·05	2·43	—
	Zn-acceptor		15	2·49	—
	S-donor		4×10^3	4·04	—
	Se-donor		3×10^3	4·16	—
	Li		0·53	1·0	—
GaSb	Intrinsic	650–700	$3{\cdot}2 \times 10^3$ (Ga)	3·15	5×10^{-13}
			$3{\cdot}4 \times 10^4$ (Sb)	3·44	1×10^{-13}
InSb	Intrinsic	450–500	0·05 (In)	1·81	5×10^{-7}
			0·05 (Sb)	1·94	3×10^{-7}
			10^5 (In)	3·85	10^{-13}
InP	Intrinsic	850–1000	7×10^{10} (P)	5·65	10^{-14}

* Values quoted here are selected from SEEGER and CHIK[44] and MADELUNG. 1964. (*Physics of III–V Compounds.* John Wiley, New York.)

Making use of Equ. 1.6 to describe the experimental results at T K we find

$$\eta > \frac{\hbar D_0^d}{a^2 k \theta_D} \exp\left(-\frac{E_D}{kT}\right) \tag{5.6a}$$

Substituting values of D_0 and E_D from Table 5.2 and the known values of a, θ_D, k and $\hbar$ we obtain a minimum concentration of $2{\cdot}5 \times 10^{-10}$ for Si and $1{\cdot}25 \times 10^{-10}$ for Ge at their respective melting temperatures.

If the value of E_D is known for a particular defect mechanism we may also derive upper limits for η by assuming that the experimentally determined diffusion coefficient results only from that particular mechanism. Using arguments similar to those above and including correlation effects, we obtain for a single vacancy mechanism

$$\boxed{\eta \leqslant \frac{8D_0^v(\text{experiment})}{a^2 f_v \nu_v} \exp\left(-\frac{E_M}{kT}\right)} \tag{5.7}$$

where f_v and E_M are respectively correlation factor and migration energy of the vacancy and ν_v is its jump frequency. We shall see later that monovacancies have migration energies of 0·33 eV in Si and 0·2 eV in Ge. Thus we expect the maximum concentrations of single vacancies to be $1{\cdot}8 \times 10^{-7}$ and $7{\cdot}4 \times 10^{-8}$ respectively at the melting points of Si and Ge.

In close-packed metals single vacancy concentrations at the melting point are in the range 10^{-5} to 10^{-3}. Such small concentrations for semiconductors must imply that the defect formation energy E_F is large since $E_D = E_F + E_M$. Alternately one must look to other mechanisms such as *ring diffusion* for an explanation: in such open lattice structures these other processes may readily take place. Theoretical predictions of defect energies are not particularly successful. Although they yield large formation energies for the single vacancy they do not predict the small migration energies which are measured experimentally. Thus both the theoretical and experimental understanding of defect properties is less soundly based in valence crystals than in the alkali halides.

Self-diffusion in III–V compounds

The III–V compounds have much in common with Si and Ge. In general the same experimental techniques, which determine the penetration profile of some radioactive tracer into the bulk crystal, are used. Table 5.2 compares values of E_D, D_0 and D (m.t.) for some III–V compounds with corresponding values in Si and Ge. The following general conclusions are apparent:

(i) the pre-exponential factors D_0 are comparable to those in Si and Ge,

(ii) Diffusion proceeds even more slowly in the III–V compounds mainly as a result of the larger values of E_D compared with Si and Ge,

(iii) The two components diffuse separately on their respective sublattices otherwise their diffusion coefficients would be equal. This follows from the crystal structure of the III–V compounds since all nearest-neighbours of a given atom belong to the other sublattice (see Fig. 5.1). Thus substitutional diffusion occurs either by direct interchange between like nearest-neighbour atoms or by an atom jumping into a vacancy located at the next nearest-neighbour site.

(iv) The values of E_D and D_0 are larger for the pentavalent atoms than for the trivalent atoms. A further consequence of these results is that the thermal equilibrium concentration of intrinsic defects in the III–V compounds should be in the same range as in the elements Si and Ge.

Impurity effects on self-diffusion

Self-diffusion in semiconductors may be modified by *n*-type or *p*-type impurities. For self-diffusion in Ge, E_D and D_0 decrease significantly in the presence of *n*-type impurities and increase in the presence of *p*-type impurities (see Table 5.2). The magnitude of the effect suggests that a defect mechanism is involved in which both the energies of defect formation and migration depend upon the Fermi level. These features are predicted by a method outlined by Seeger and Chik.[44]

Suppose that the defects which promote diffusion introduce an acceptor level E_i into the band gap of a degenerate semiconductor. Let the defect concentrations be η^i and η^d in the intrinsic and doped materials respectively, where the Fermi levels are ζ^i and ζ^d. At equilibrium the concentrations take the form,

$$\eta = \exp(S_F/kT) \exp -(E_F/kT) \times F(t) \tag{5.8}$$

S_F and E_F are respectively the entropy and energy of defect formation. The function $F(t)$ which represents the equilibrium between charged and uncharged defects is given[45] by $F(t) = 2 + \exp\{(\zeta - E_i)/kT\}$. Thus the ratio of defect concentrations in intrinsic and doped material is,

$$\begin{aligned} \frac{\eta^d}{\eta^i} &= \frac{2 + \exp\{(\zeta^d - E_i)/kT\}}{2 + \exp\{(\zeta^i - E_i)/kT\}} \\ &= \frac{2 + \epsilon(T)\exp\{(\zeta^d - \zeta^i)/kT\}}{2 + \epsilon(T)} \end{aligned} \tag{5.9}$$

where $\epsilon(T) = \exp\{(\zeta^i - E_i)/kT\}$. The electron concentration in the conduction band is related to the function $\exp\{(\zeta^d - \zeta^i)/kT\}$ by,

$$n_e = n_i \exp\{(\zeta^d - \zeta^i)/kT\} \tag{5.10}$$

where n_i is the electron concentration in the conduction band under intrinsic conditions. In *n*-type material the requirement of charge neutrality leads to $n_e \approx N_d$ where N_d is the density of impurity donors. Substituting into Equ. 5.9 from Equ. 5.10 gives

$$\boxed{\frac{\eta^d}{\eta^i} = \frac{2 + \epsilon(T)\dfrac{N_d}{n_i}}{2 + \epsilon(T)}} \tag{5.11a}$$

Similarly in a *p*-doped solid we find

$$\boxed{\frac{\eta^d}{\eta^i} = \frac{2 + \epsilon(T)\dfrac{n_i}{N_a}}{2 + \epsilon(T)}} \tag{5.11b}$$

Thus at constant temperature T the concentration of defects is increased by p-type impurities. Diffusion by an acceptor defect is consequently slower in p-type material than in n-type or intrinsic material. The reverse situation obtains if the intrinsic defect acts as a donor.

Impurity diffusion

In the last two decades many papers have been published on the diffusion of impurities in semiconductors. Two main experimental methods are used: radioactive tracer techniques remain quite common. However, the p-n junction method finds wider application to the study of electrically active impurities. This latter method measures diffusion through a host matrix doped with carriers of opposite charge to those introduced by the impurity. Thus the diffusion coefficients of an electrically active impurity migrating through a uniformly doped matrix are studied. If the radioactive tracer and p-n junction methods are used together then the contributions by charged and uncharged impurities to the total conductivity may be assessed.

Diffusive impurities are divided into two categories; either they are *fast diffusors* or *slow diffusors* according to the magnitude of their diffusivities. In Si and Ge impurities from Groups III and V of the Periodic Table are usually slow diffusors. The mechanisms of slow diffusion are expected to be related to those of self-diffusion. Table 5.2 shows that slow diffusors have diffusion coefficients which are up to 10^2 times higher than the self-diffusion coefficients. In Ge Group III acceptors diffuse about 100 times slower than Group V donors, mainly as a result of E_D being increased for acceptors and decreased for donors. Although the pre-exponential factors (D_0) may be large for Group III elements this is more than compensated by the exponential term. In Si both Group III and Group V elements have diffusion coefficients closely similar to those for self-diffusion, with the donor elements diffusing slightly the more slowly. The results given in Table 5.2 are in no way exhaustive, nor do they indicate the wide discrepancies between results of different authors caused by the different experimental conditions. The magnitude of doping and the surface concentration of impurities seem to be particularly influential in respect of diffusion coefficients. This may result from the different diffusion constants of ionized impurities creating an internal electric field in the semiconductor which affects the impurity distribution.

The most important fast diffusors are from Groups I and VIII: on account of their small atomic size they appear to migrate by an interstitial mechanism. A common feature of fast diffusors is that the experimental results are not described by Equ. 1.6 with E_D and D_0 being temperature independent quantities: the crystalline perfection of the specimens is usually found to be particularly important. The mechanism of diffusion of elements from Groups I and VIII in Si and Ge is rather more complicated than for slow diffusors.

These elements dissolve either substitutionally or interstitially and consequently they diffuse by both vacancy and interstitial mechanisms. These impurities may also act as either donors or acceptors. Where both interstitial and substitutional mechanisms can be differentiated, it is found that the activation energy for interstitial diffusion is much lower than that for substitutional diffusion. The mechanism for Li, however, seems to be unequivocally determined as being interstitial only. In general, however, the interpretation of experimental data for fast diffusors requires that more than one mechanism be taken into account.

GaAs has been most intensively studied of the III–V compounds. It is found that both D_0 and E_D of acceptors on the trivalent sublattice are decreased very significantly relative to pure specimens. Donors on the pentavalent sublattice also have diffusion coefficients smaller than those for self-diffusion. According to the distinctions noted above for Si and Ge these elements in GaAs must be regarded as slow diffusors. Elements from Groups I and VIII are again very fast diffusors, having very small activation energies compared with self-diffusion. In character with fast diffusors in other systems Li does not obey Fick's laws of diffusion.[45] This is connected with the formation of complexes between interstitial Li^+ ions and substitutional Li^{2+} ions, which act together as a single acceptor.

5.3 Radiation-induced defects in silicon

The earliest electron paramagnetic resonance studies of irradiation damage in silicon were reported for neutron irradiated crystals. A number of the reported spectra were attributed to intrinsic defects since their production was apparently impurity independent. However, neutron damage is undoubtedly very complex. Watkins and his co-workers[46] sought some simplification by studying electron irradiation damage. Historically the structures of several vacancy/impurity complexes were recognized in electron irradiated samples before the single vacancy had been specifically identified. This was because the earliest studies of electron irradiation were conducted on samples irradiated near room temperature, where the lattice vacancy is mobile. These later studies have led to an assignment of the earlier spectra based upon clusters of single vacancies detected in the electron irradiated crystals.[10]

The structure of the single vacancy

When a single lattice vacancy is created in the Si lattice four (sp^3) hybrid bonds are broken which leaves the four "dangling" bonds shown in Fig. 5.5a pointing into the vacancy from the tetrahedrally disposed atoms. Each broken bond contains one electron which may occupy a lower energy configuration by forming new covalent bonds across the vacancy with the broken bonds on the other neighbouring atoms. The wavefunctions of the

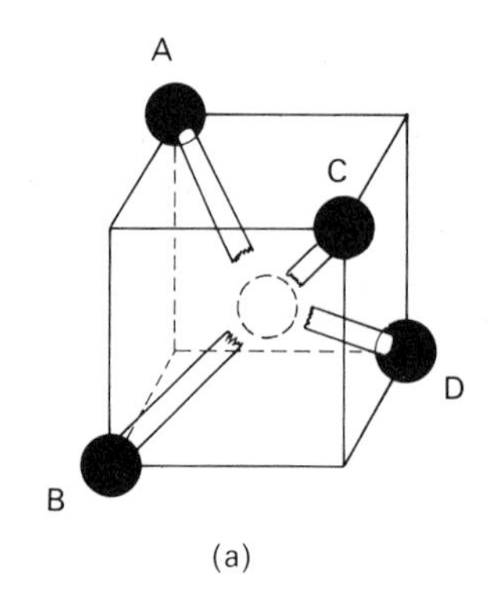

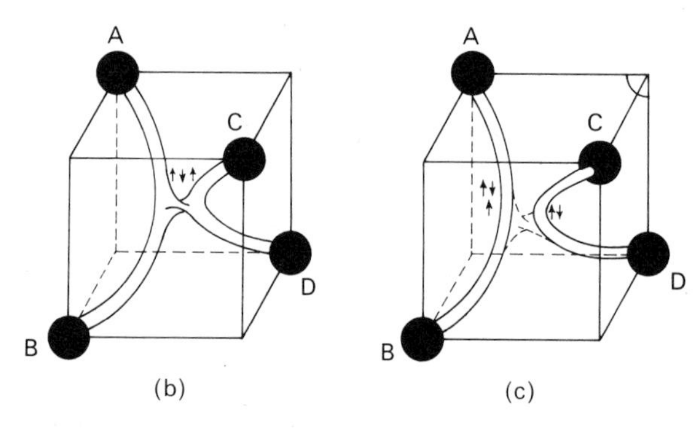

Fig. 5.5 Single vacancy configurations in silicon. Showing (a) the four "dangling" bonds around the vacant lattice site, (b) the V^+-three electron configuration neglecting distortion and (c) the V^--five electron configuration neglecting distortion. (Adapted from WATKINS, G. D. 1968. *Radiation Damage in Semiconductors.* Plenum Press, New York.)

electrons in these bonding orbitals are constructed by taking a linear combination of the broken bond orbitals on each of the four neighbouring atoms. The lowest energy configuration corresponds to the totally symmetric wave function,

$$\psi = \psi_A + \psi_B + \psi_C + \psi_D \tag{5.12}$$

since this represents bonding between all four atoms (Fig. 5.5b). Three alternative and degenerate wave functions exist at a higher energy (Fig. 5.6). When only three electrons are distributed over these orbitals (i.e. the vacancy is positively charged, V^+), two electrons will pair off in the bonding orbital with spins mutually opposed. The third electron will enter one of the higher energy antibonding orbitals (Fig. 5.6). The single vacancy in this charge state will be paramagnetic with electron spin $S = \frac{1}{2}$. If five electrons are localized in these orbitals a slightly different configuration (Fig. 5.5c) results, although the vacancy is again paramagnetic.

Watkins[46] recorded the ESR spectrum of the vacancy at 2 K to 20 K, in *p*-type silicon after irradiation below 40 K with 1·5 MeV electrons. The spectrum consists of three equally intense lines, each at the centre of four pairs of very weak lines. The positions of the principal lines are determined by the *g*-tensor of the spectrum which may be written as,

$$g(\theta) = (g_{\perp}^2 \sin^2 \theta + g_{\parallel}^2 \cos^2 \theta)^{\frac{1}{2}} \tag{5.13}$$

θ being the angle between the $\langle 100 \rangle$ axis and the static magnetic field. The measured *g*-values are $g_{\perp} = 1{\cdot}9989$, $g_{\parallel} = 2{\cdot}0087$. The weaker lines are

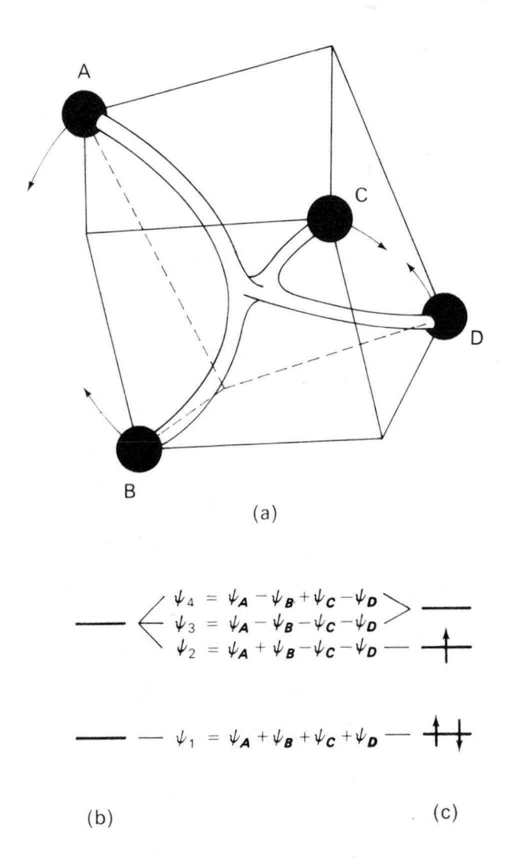

Fig. 5.6 Showing (a) the axially symmetric distortion of the V^+ configuration in silicon (b) the one electron molecular orbitals before distortion (c) the molecular orbitals after distortion. (After WATKINS, G. D. 1968. *Radiation Damage in Semiconductors*. Plenum Press, New York.)

caused by magnetic hyperfine interaction with the 4·7% abundant ^{29}Si nuclei ($I = \frac{1}{2}$) at four different Si sites around the vacancy. The positions of these lines in magnetic field units are,

$$\boxed{H = H(\theta) + m_I(A + B\{3\cos^2\psi - 1\})} \qquad (5.14)$$

where $H(\theta) = h\nu/g(\theta)$, ψ is the angle between the hyperfine axis and the static magnetic field, $A = 34{\cdot}5G$ and $B = 4{\cdot}7G$. Detailed analysis shows that the hyperfine axis is tilted about 7° away from the $\langle 111 \rangle$ axis toward the $\langle 100 \rangle$ axis associated with $g_{\parallel}$. As discussed in Chapter 4 the hyperfine structure is a measure of the spatial extent of the electronic wavefunction. For the V^+ state, to which this spectrum is attributed, the hyperfine structure is consistent with the vacancy-centred orbitals being approximately 22% 3*s* character and 78% 3*p* character. Furthermore over 60% of the total electronic wave function is distributed over the four nearest neighbour atoms.

The axially symmetric *g*-tensor and the tilting of the hyperfine axis away from the $\langle 111 \rangle$ direction are evidence that the defect is distorted, the four surrounding atoms being displaced from their normal lattice points. A possible distorted configuration which is axially symmetric about a $\langle 100 \rangle$ direction is shown in Fig. 5.6. There are three equivalent orientations of this distortion, which are normally equally populated with defects. Each orientation produces one of the 3 principal lines in the ESR spectrum (neglecting the hyperfine structure). However, the energy barrier for reorientation among these three distortion configurations is small and thermally activated reorientation should be observed at low temperatures. In the temperature range 14 to 20 K significant broadening of the spectral lines is observed which arises when a defect with an anisotropic *g*-tensor jumps from one orientation in the lattice to another, since its *g*-value changes abruptly with each orientation. Thus defects will alternately be brought into and out of resonance as they randomly reorientate. Such motion will give rise to further broadening with increasing temperature since the rate of reorientation increases. Watkins[46] finds an activation energy for reorientation in the range 0·01 to 0·02 eV with a reorientation rate of order $\tau = 10^{-8}$ sec. These values are confirmed by studying the reorientation as a function of uniaxially applied stress. Since the defects themselves can reorientate thermally they are able to seek a preferred orientation in a crystal subjected to a uniaxial stress. A typical result is shown in Fig. 5.7. The relative intensities of the lines change when a uniaxial $\langle 100 \rangle$ stress is applied, indicating that a preferential alignment of the defect has occurred. It should be emphasized that this reorientation is "electronic," i.e. bond-switching is involved not atomic movements of the molecular unit.

Although the ESR spectrum strongly suggests the single vacancy model, the motional effects are quite emphatic since the reorientation times and small activation energies must result from quantum mechanical effects rather

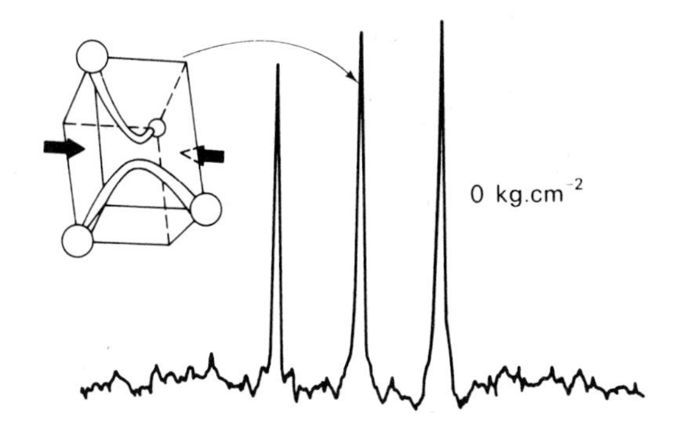

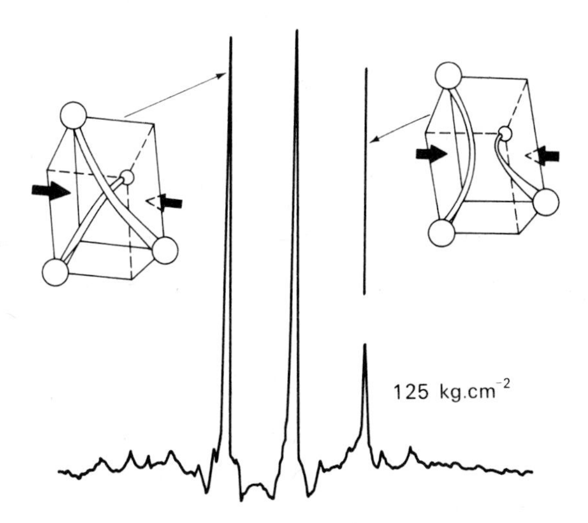

Fig. 5.7 The change in the ESR spectrum of the V^+-centre in Silicon as a result of re-orientation under a [100] compressional stress. The insets show the defect orientation corresponding to each multiplet. (After WATKINS. 1963. *J. Phys. Soc. Jap.* **18** (Supp)., 22.)

than from atomic rearrangement. The vacancies must also be isolated from other lattice imperfections, otherwise they would be influenced by distortions associated with neighbouring defects. The response by vacancies to applied stresses implies an upper limit for the average strain at each vacancy of only 2×10^{-5}. This is characteristic of the internal strains in unirradiated crystals. Furthermore, since the hyperfine interaction involves all four neighbouring atoms the vacancy must be in the positively charged V^+-state which is unstable as long as the Fermi level is locked to the shallow acceptor levels. Apparently the neutral vacancy ionizes as a donor when the Fermi level is below $E_v + 0{\cdot}05$ eV. The vacancy becomes negatively charged when the Fermi level is raised by further electron irradiation to $\zeta = E_v + 0{\cdot}3$ eV.* The V^--vacancy configuration is also paramagnetic with $S = \frac{1}{2}$. A small g-shift relative to the g-value for free electrons is observed in the ESR spectrum, as was the case for the V^+-centre. The observed hyperfine structure, however, differs appreciably and it confirms that the unpaired electron resides on only two of the surrounding nuclei[10,46] This is consistent with the notions on bond formation in the V^--centre; in effect two bonds are formed, one between atoms A and B, and one between atoms C and D. The five electrons are distributed over these bonds leaving the odd unpaired electron resident in an anti-bonding orbital as shown in Fig. 5.5c. Since these orbitals extend between only two of the neighbouring Si atoms hyperfine interaction is observed only with these two nuclides. The spectra associated with V^+ and V^- configurations are not observed in low resistivity n-type material, where the V^{2-} state is thought to be stable.

Divacancies in silicon

Divacancies are also primary products of electron irradiation in silicon although the efficiency of divacancy production is less than 5% that of single vacancies. The structure of this primary defect is shown in Fig. 5.8 to involve two vacancies on the adjacent sites A and B in the lattice. Around these vacancies are six "broken" bonds, one for each of the six neighbouring atoms. The new bonds formed between these neighbours include two "bent" pair bonds between atoms 2 and 3 and between atoms 5 and 6, and an "extended molecular bond" between atoms at positions 1 and 4. Twelve possible orientations of the divacancy result from the four-fold equivalence of the $\langle 111 \rangle$ directions which join together the two vacancies. Each orientation of the defect axis has three equivalent directions for the extended molecular bond (e.g. in Fig. 5.8 this bond may alternatively be formed between atoms at positions 2 and 5 or at positions 3 and 6). These extended molecular bonds are relatively weak and lie to higher energies than the bent pair bonds. The defect is paramagnetic with $S = \frac{1}{2}$ when either one or three electrons occupy the extended orbital, the charge states being respectively positive (V_2^+) and negative (V_2^-). Both of these configurations have been identified by ESR.

The ESR spectra of the divacancy are similar to those of the single vacancy spectra, the principal components being attended by weaker multiplets due to magnetic hyperfine interaction with neighbouring ^{29}Si nuclei.

* ζ refers to the Fermi level whilst E_c and E_v are respectively the energies of electrons at the top of the valence band and the bottom of the conduction band.

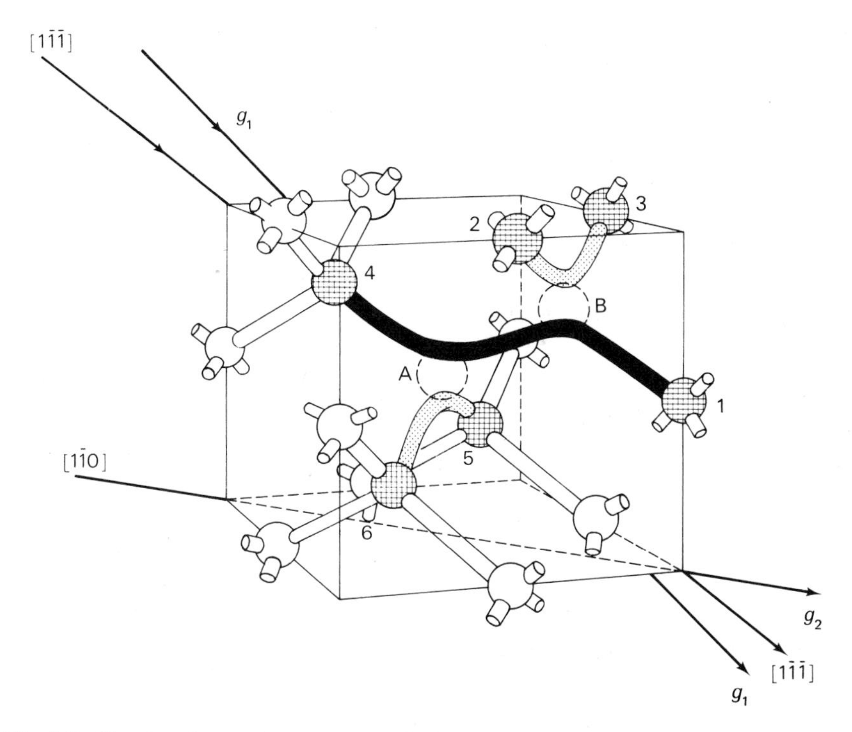

Fig. 5.8 Showing the atomic structure and bond configuration in the divacancy in silicon. The effects of distortion are neglected. In the V_2^+ configuration a single electron is localized in the extended bond between atoms 1 and 4. The V_2^- configuration has three electrons in this bond. (After WATKINS and CORBETT. 1965. *Phys. Rev.* A138 A543.)

The line positions are given by Equ. 5.14, suitably modified to take account of the different symmetry of the g-tensor. For the divacancy $g(\theta)$ is replaced by

$$g(\theta, \phi) = (g_1^2 \sin^2 \theta \cos^2 \phi + g_2^2 \sin^2 \theta \sin^2 \phi + g_3^2 \cos^2 \phi)^{\frac{1}{2}} \quad (5.15)$$

where θ and ϕ are the polar and azimuthal angles relative to the magnetic field direction. g_1, g_2 and g_3 are closely similar to the g-value for free electrons and are directed approximately along (for example) the [11$\bar{1}$], [011] and [21$\bar{1}$] axes. The intensity of the hyperfine components relative to the principal lines is just twice the isotopic abundance of ^{29}Si. This indicates the two-fold equivalence of the sites which accommodate a ^{29}Si nucleus. The hyperfine interaction is axially symmetric about that $\langle 111 \rangle$ axis parallel to the extended molecular axis. A comparison of the parameters which describe the ESR spectra of the divacancy with those of the single vacancy and other defects is

Table 5.3 The properties of vacancy, vacancy pair and vacancy-impurity pairs in silicon (determined by ESR)

Defect configuration and spin S	g-value	Hyperfine constants ($\times 10^{-4}$ cm^{-1})		Hyperfine axes + no. of equivalent atoms
		^{29}Si	Impurity	
V^+ $S = \frac{1}{2}$	$g_\parallel = 2{\cdot}0087$ $g_\perp = 1{\cdot}9989$	$A = -34{\cdot}5$ $B = -4{\cdot}7$	— —	⟨111⟩ 4
V^- $S = \frac{1}{2}$	$g_1 = 2{\cdot}0151$ $g_2 = 2{\cdot}0028$ $g_3 = 2{\cdot}0038$	$A = -118{\cdot}9$ $B = -7{\cdot}2$	— —	⟨111⟩ 2
V_2^+ $S = \frac{1}{2}$	$g_1 = 2{\cdot}0004$ $g_2 = 2{\cdot}0020$ $g_3 = 2{\cdot}0041$	$A = -49{\cdot}3$ $B = -9{\cdot}3$	— —	⟨111⟩ 2
V_2^- $S = \frac{1}{2}$	$g_1 = 2{\cdot}0012$ $g_2 = 2{\cdot}0135$ $g_3 = 2{\cdot}0150$	$A = -64$ $B = -10$	— —	⟨111⟩ 2
$(V + O_i)^-$ $S = \frac{1}{2}$	$g_1 = 2{\cdot}0092$ $g_2 = 2{\cdot}0026$ $g_3 = 2{\cdot}0033$	$A = -136{\cdot}8$ $B = -8{\cdot}1$	— —	⟨111⟩ 2
$(V + P_s)$ $S = \frac{1}{2}$	$g_1 = 2{\cdot}0005$ $g_2 = 2{\cdot}0112$ $g_3 = 2{\cdot}0096$	$A = -115{\cdot}7$ $B = -17{\cdot}2$	$a = +9{\cdot}32$ $b = +0{\cdot}63$	⟨111⟩ 1(^{29}Si) 1(^{31}P)
$(V + Al_s)^-$ $S = 1$	$g_\parallel = 2{\cdot}0136$ $g_\perp = 2{\cdot}0085$ $D = 0{\cdot}414$ $\times 10^{-4}$ cm^{-1}	$A = -31{\cdot}4$ $B = -2{\cdot}9$	$a = +15{\cdot}29$ $b = -0{\cdot}14$	⟨111⟩ 3(^{29}Si) 1(^{27}Al)

The values are taken from Tables given in ref. 10. The superscripts indicate the charge state of the defect and subscripts indicate whether the impurity is interstitial (*i*) or substitutional (*s*). The activation energies for diffusion refer to the neutral state of the defect.

given in Table 5.3. The g-shifts and the hyperfine constants are slightly larger for the negatively charged configuration than for the positively charged configuration. The hyperfine interactions are consistent with 60% of the wave functions being located on the two principal sites. A further 20% of the wave functions is localized on the four remaining sites which neighbour the divacancy. The remaining wave function is presumably spread out over more distant sites in the crystal.

Motional effects at higher temperatures are also observed in the ESR spectra of the divacancy. The lifetime broadening and motional averaging of the spectra over the temperature 15 to 100 K are consistent with the bond reorientation occurring with lifetimes which are slightly different for the two charge states of the divacancy. The experimental results of this study are fitted to Equ. 2.15 in Fig. 5.9. Since the results shown on this plot of $\log_e \tau$ versus T^{-1} span fourteen decades in the lifetimes, the activation energies and pre-exponential factors are determined with high accuracy. The actual magnitudes of τ_0 and E indicate that the molecular bond reorientates more easily in the V_2^- configuration than in the V_2^+ configuration. Figure 5.9 also contains values of τ determined from experiments involving the application of a uniaxial stress to the crystal. The applied stress distorts the crystal symmetry and the various orientations of the bond axis are no longer equivalent. The defects are free to reorientate and take up the lowest energy configuration in the strained crystal. Thus the characteristic re-orientation times for the spectrum under stress correspond to those involved in the linewidth studies. The sense of the realignment under stress is such that the preferred defect configuration is that in which the silicon atoms joined by "bent" pairs bonds are pushed closer together. The magnitude of the alignment is consistent with a lowering of the energy per electron in the pair bond of approximately 6 eV/unit strain and 8 eV/unit strain for the V_2^+ and V_2^- configurations respectively.

At higher temperatures (270 to 470 K) the vacancy-vacancy axis also re-aligns under stress: this axis remains fixed at low temperature because the atomic rearrangements require much more energy than bond reorientation. Watkins and Corbett[47] applied a uniaxial stress to the crystal at elevated temperatures prior to cooling to 20 K for ESR measurement of the degree of alignment. From Fig. 5.8 it is evident that the divacancy reorientates when atom 6 jumps to vacancy B via vacancy A. Since the divacancy has acquired a new axis and been displaced by one lattice spacing, it has apparently migrated or diffused away from its original site. By applying first order kinetics to their results Watkins and Corbett determined an activation energy for divacancy diffusion of 1·3 eV. During this reorientation the two vacancies are separated. Since no loss of divacancies is observed at 270 to 470 K there is a substantial binding energy between the two separated vacancies. In fact the binding energy of the divacancy must be greater than the difference between the activation energies for diffusion of the divacancy and single vacancy, i.e. 1·3 − 0·3 eV = 1·0 eV. This is confirmed by measurements at 470 to 670 K, where the divacancy anneals out. Watkins and Corbett suggest that the binding energy of the divacancy must be $\approx$1·6 eV.

The two charge states of the divacancy are stable under quite different experimental conditions. This is readily apparent in the energy level diagram shown in Fig. 5.10, which is constructed using the linear combination of atomic orbitals (LCAO) scheme discussed for the single vacancy. According to this scheme the charge on the divacancy is determined by the filling of the ψ_C and ψ_D levels in Fig. 5.10. With one or three electrons in these levels the divacancy has a net charge of +1 and −1 respectively. The positively charged state is observed in p-type silicon. However it is an acceptor level in the band gap and may be optically bleached by excitation of a hole from

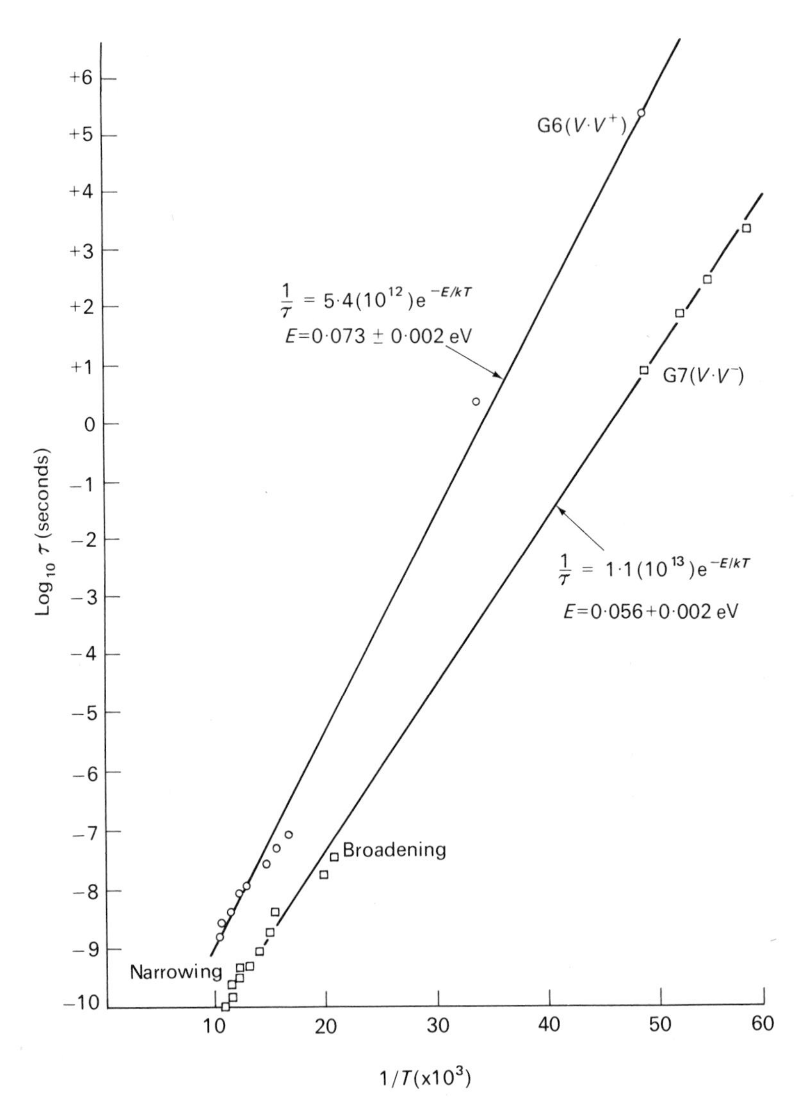

Fig. 5.9 Reorientation times of the distorted configurations of divacancies in Si as a function of temperature. (After WATKINS and CORBETT. 1965. *Phys. Rev.* A138 A543.)

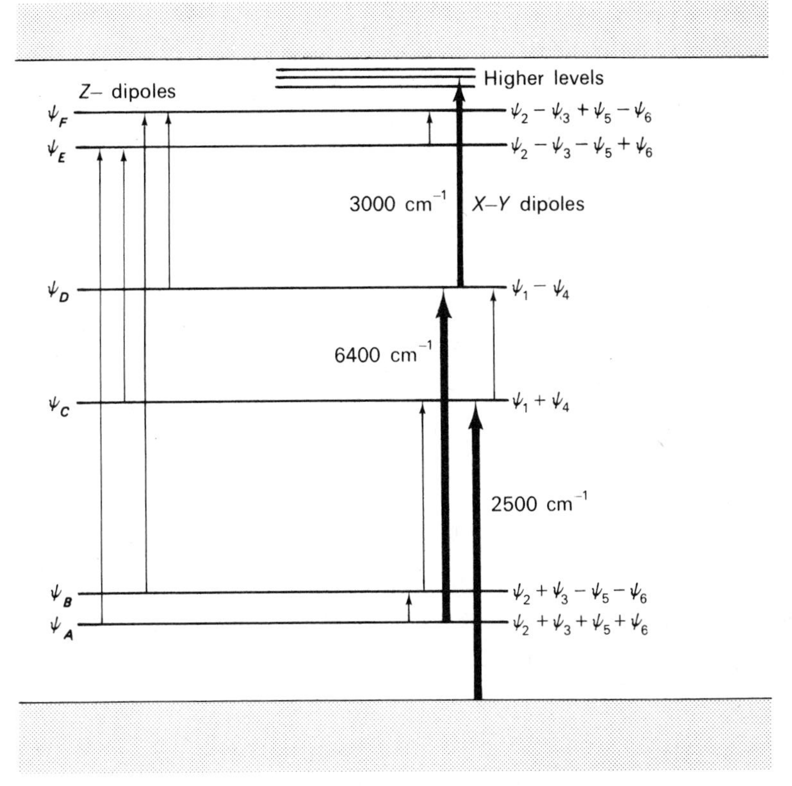

Fig. 5.10 Schematic representation of the molecular orbital energy levels of the divacancy. The allowed electric dipole transitions are indicated ↑: observed transitions ↑ are the most probable bands consistent with this scheme. (Modified from CHENG *et al.* 1966. *Phys. Rev.* **152,** 761.)

the defect to the valence band edge. This process occurs at photon energies of $\approx$0·25 eV (i.e. $\lambda^{-1} = 2\,000$ cm^{-1}). Consequently the V_2^+ configuration is stable only when $\zeta \leqslant E_v + 0{\cdot}25$ eV. The divacancy is neutral when two electrons are present in the extended molecular orbital. The V_2^- configuration is observed whenever the Fermi level is between the intrinsic Fermi level and $E_c - 0{\cdot}4$ eV. Above $E_c - 0{\cdot}4$ eV the doubly negative charge state is stable.

Figure 5.10 shows also that there are numerous allowed optical transitions of the divacancy in silicon which should be observed in the infra-red spectrum. Some typical infra-red spectra of n-type and p-type silicon are shown in Fig. 5.11. The transitions at 2 500 cm^{-1} (3·9 μ), 3 000 cm^{-1} (3·3 μ) and 6 400 cm^{-1} (1·8 μ) are due to some of the transitions at the divacancy shown in Fig. 5.10. Other possible transitions have not been identified.

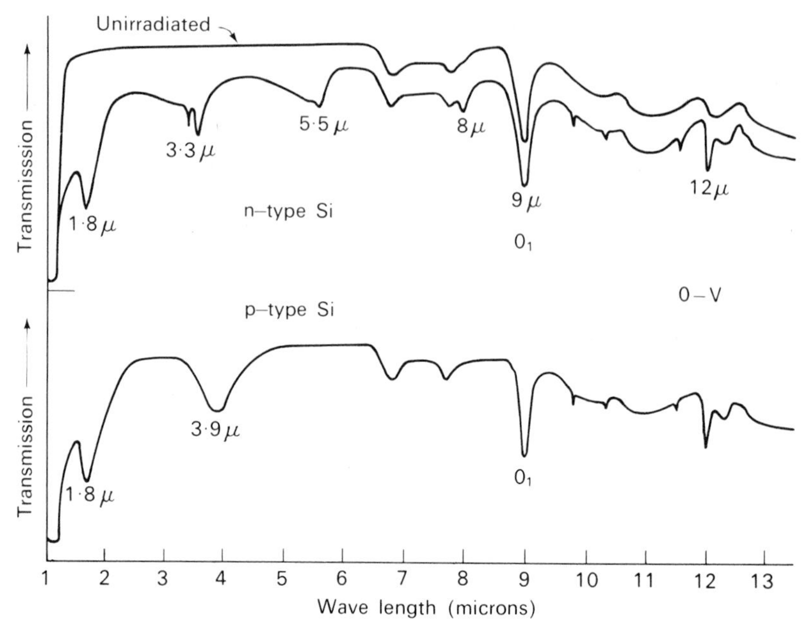

Fig. 5.11 Schematic diagram of radiation-induced infra-red absorption bands in both n-type and p-type silicon in the wavelength range 1 to 13 μ. (After CHENG *et al.* 1966. *Phys. Rev.* **152**, 761.)

Vacancy-impurity complexes

Although early studies of radiation damage in silicon identified specific electrical, optical and mechanical property changes arising from the presence of a particular defect, it was not possible to identify that defect. As we have already indicated the many careful electron spin resonance studies by Watkins and his associates[46,47] have to a large extent rectified this situation. This discussion so far, however, has taken little account of the chronological sequence of events, since in actuality the importance of impurities as trapping centres for mobile vacancies was recognized long before the single vacancy. This came about mainly because radiation damage was introduced into the crystal at room temperature, where a variety of vacancy-impurity centres are produced. The most important impurities appear to be O, P, Al, Ga and As. The structure of the vacancy-oxygen, vacancy-phosphorus and vacancy-aluminium pairs are shown in Fig. 5.12. The essential difference between these defect-pairs is that the oxygen atom is interstitial being very close to the vacancy, whereas phosphorus and aluminium each occupy a substitutional site. The appropriate electron spin resonance parameters which were essential in revealing the defect structures are reported in Table 5.3.

The oxygen-vacancy structure involves an interstitial oxygen atom which forms bridging bonds to two silicon atoms neighbouring the vacancy, i.e. a Si–O–Si bond is formed. A covalent bond is established between the remaining two silicon atoms. The defect is paramagnetic only if the Fermi level $\xi > E_c - 0{\cdot}17$ eV. There are then three electrons in the molecular bond giving a resultant spin $S = \frac{1}{2}$, since two electrons have their spins aligned antiparallel. It has also been demonstrated that the infra-red band at

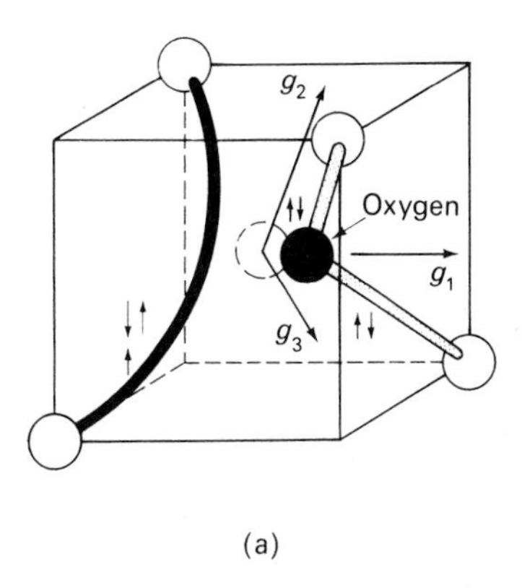

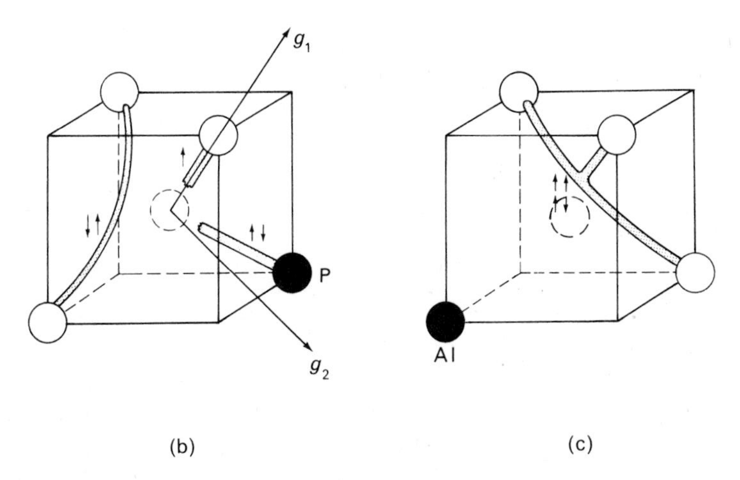

Fig. 5.12 Showing the atomic configuration of (a) (vacancy + O_i)$^-$ pair, (b) (vacancy + P) pair and (c) (vacancy + Al_s)$^-$ pair in silicon. (●-impurity, ○ vacancy). Effects of distortion are neglected. Superscripts indicate the charge state of the paramagnetic configuration, small arrows indicate the distribution of electrons and their relative spin orientation in the newly formed bonds. (Modified from WATKINS, G. D. 1968. *Radiation Damage in Semiconductors*. Accd. Press, New York.)

8300 cm^{-1} (12 μ) in Fig. 5.11 is due to antisymmetric stretching vibrations of the Si–O–Si bond. Reorientation of bonds between the atoms around the oxygen interstitial has been shown to proceed with an activation energy of $0{\cdot}38 \pm 0{\cdot}04$ eV. It should be pointed out that this last process probably involves motion in which one Si–O bond is broken when pivoting around the other.

Perhaps the most important aspect of the recognition of the vacancy-oxygen centre, however, is that it permitted accurate estimates to be made of the migration energy of the isolated vacancy. Single vacancies are produced by electron-irradiation only below $\approx$40 K. Above 40 K the vacancy is mobile and is trapped by impurity atoms. Watkins working at 130 to 160 K measured the decay of the vacancy during the formation of the oxygen-vacancy centre in crystals containing low and high oxygen contents. He observed a simple exponential process with an activation energy of 0·33 eV in p-type silicon. In low oxygen content crystals, the decay is much slower due to the vacancy diffusing further before being trapped. This measurement corresponds to the activation energy for neutral vacancy diffusion. In n-type material the vacancy migrates with an activation energy of 0·18 eV. Since the activation for self-diffusion is very high ($E_D \approx 4{\cdot}86$ eV) the results on p-type silicon imply that the vacancy formation energy is also high ($E_F \approx 4{\cdot}5$ eV).

In phosphorus doped crystals the vacancy is trapped by phosphorus atoms. The phosphorus-vacancy pair shown in Fig. 5.12b is paramagnetic ($S = \frac{1}{2}$) when the Fermi level is below $E_c - 0{\cdot}4$ eV. The substitutional phosphorus atom then has paired electrons in a "dangling" bond pointing into the vacancy, while two of the three neighbouring silicon atoms are covalently bonded together. An unpaired electron resides in a bond on the remaining Si neighbour. The ESR parameters of this defect are also given in Table 5.3.

The activation energy for diffusion of the vacancy-phosphorus pair ($\approx$0·93 eV) is a most important quantity, since the diffusion process first involves the vacancy migrating away from the phosphorus atom. Thus the difference between the activation energies for diffusion of the phosphorus-vacancy pair and the single vacancy gives a lower limit of 0·6 eV to the binding energy between the separated phosphorus atom and vacancy. Since no loss occurs in the vacancy-phosphorus pair concentration in the range 250 to 270 K this is not the total binding energy, and Watkins and Corbett[47] have argued that the binding energy between the phosphorus atom and a vacancy in the third nearest neighbour site is approximately 0·28 eV. This binding energy effectively lowers the activation energy for phosphorus diffusion relative to self-diffusion. Consequently from the activation energy for phosphorus diffusion $E_D(P) = 3{\cdot}88$ eV, we estimate that the activation energy for self-diffusion is $E_D = 3{\cdot}88 + 0{\cdot}28 = 4{\cdot}16$ eV. The corresponding formation energy of the lattice vacancy is then $E_F = 3{\cdot}83$ eV. Self-diffusion measurements on Si are not particularly consistent between different workers since such measurements are known to be particularly difficult. It seems probable therefore, that the vacancy formation energy is nearer this latter value of $E_F = 3{\cdot}83$ eV than the earlier value.

Perhaps the most significant aspect of these studies is that they give relatively accurate determinations of quantities related to diffusion processes in semiconductors. Furthermore these measurements are highly specific since they are concerned with a particular diffusion mechanism, which is not usually the case in bulk diffusion measurements. They also give detailed information about interactions of lattice vacancies with electrically active *p*- and *n*-type impurities as well as with other vacancies. The examples discussed here are but a fraction of those available to us[10,46] In general, radiation damage even with 1·5 MeV electrons is a very complex process. Many different defect structures are formed depending upon the nature of the impurities, the Fermi level and other factors. Nevertheless, the examples we have chosen demonstrate that the interactions of the vacancy with both donors and acceptors are consistent with its acting as an acceptor in *n*-type material and a donor in *p*-type material. These interactions may be viewed as further manifestations of the Coulomb attraction between unlike charges.

Interstitial defects in silicon

Isolated interstitials or small clusters of interstitial Si atoms have never been observed after irradiation. This is a rather curious result since, as we discussed in Chapter 2, equal numbers of separated interstitials and vacancies should result from energetic particle irradiation. Of course the ground state of the interstitial species may be diamagnetic, in which case E.P.R. measurements are of little value. Theoretical work is of little assistance since the electronic structure of the interstitial in Si is difficult to predict, although it should be similar to the interstitial in diamond. Yamaguchi[48] has suggested that the ground state of the neutral interstitial carbon atom in diamond is diamagnetic. However, a spin triplet state should occur approximately 0·1 to 0·4 eV above the ground state and could give rise to ESR at high temperatures. Other charge configurations of the interstitial (e.g. I^+ and I^-) should be paramagnetic but have not been observed. If any of these configurations are present in irradiated Si they are not observed for one reason or another. The extreme mobility of the interstitial Si is a complicating factor, and it is probable that such imperfections are only stable at exceedingly low temperatures where the excited triplet states are likely to be empty.

A number of pieces of indirect evidence concerning the fate of the Si interstitials have been reported. After electron irradiation at 4 K Watkins[46] identified Al^{2+} interstitials in *p*-type Si containing aluminium. These interstitials were produced at the same rate as V^+-centres in the same sample. This suggests that the Si interstitials produced during the electron irradiation are mobile at the irradiation temperature and migrate through the lattice until they are trapped at and change places with the substitutional Al^{2+}

atoms. Thus the presence of Al^{2+} interstitials reflects that the irradiation produces Si interstitials which migrate with an activation energy of $\approx$ 0eV in p-type Si. On annealing at higher temperatures the Al^{2+} interstitial associates with a substitutional Al, to form a defect pair in the configuration $(Al_i^{2+} + Al_s^-)$. ESR also detects other impurity interstitial species including $(Ga_i^{2+} + Ga_s^-)$ and a boron configuration of undetermined origin.

In n-type material the response to a unaxial stress of divacancies produced at low temperature differs from that of divacancies formed at elevated temperature. This behaviour persists on annealing the material until at 140 K about 80% of the divacancies disappear. Watkins interpreted this result as being due to interstitials, which are frozen into the lattice at low temperature as a consequence of their being formed near divacancies, becoming mobile at 140 K and partially annihilating the divacancies. Under these circumstances, he argues, the activation energy for interstitial migration in n-type material is of the same order of magnitude as the vacancy migration energy. This contradicts the infra-red absorption evidence of Whan,[49] who studied the interchange between the Si interstitial and the oxygen in the $(V + O_i)$ centre during electron irradiation at 80 K. The interstitial appears to be already mobile at 80 K, implying a relatively small migration energy. It is of course possible that the interstitials migrate to large clusters of interstitials, a situation which frequently obtains in metals and insulators.

Some general comments

We have considered a number of specific defect configurations in n-type and p-type silicon. Whether or not the ESR spectra of these defects are observed depends strongly upon the position of the Fermi level. This was particularly evident in the discussion of the vacancy and divacancy, both of which may introduce acceptor levels or donor levels into the band gap. The charge states of these two primary defects are compared with the levels introduced by other defects in Fig. 5.13. This schematic representation of the electrical levels introduced by point imperfections also indicates the condition under which we might expect to observe the appropriate ESR spectrum. There are, in fact, a host of other levels introduced by other defects, and many of these have been detected using electrical conductivity, photoconductivity and Hall effect measurements. It is inappropriate to go into further details here, except to say that very careful studies are necessary to recognize the defects responsible for all the levels observed.

The defect structure of irradiated germanium is less well understood despite the first irradiation studies preceding those on silicon. This is mainly due to experimental difficulties in applying ESR to the study of Ge. However, theoretical studies, which are not necessarily a good guide, strongly suggest smaller formation energies, binding energies and migration energies in Ge than in Si. Actually, the observation that E_M for the vacancy in Si was only

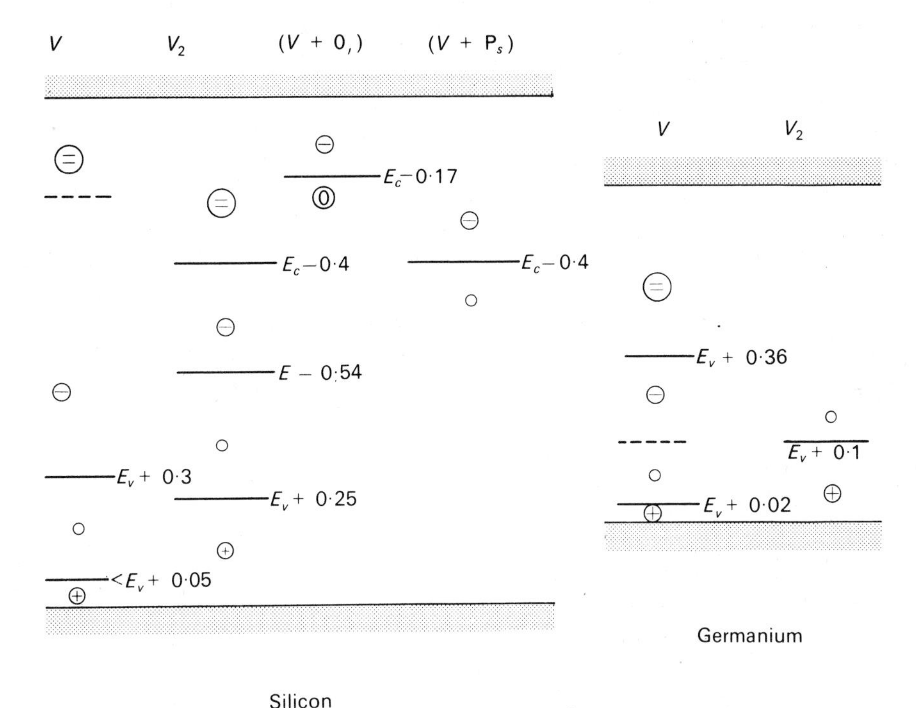

Fig. 5.13 Energy levels associated with various defects in Si as determined by ESR measurements 10. The level positions are only approximately known unless otherwise indicated. The levels indicated for germanium are taken from reference 44 and require experimental justification.

0·33 eV led to the first serious doubts about the interpretation of earlier quenching experiments on Ge in which the vacancy migration energy was suggested to be 1·0 to 1·3 eV. Subsequent infra-red measurements on oxygen-doped Ge showed that the vacancy migrated with an activation energy of only $\approx$0·2 eV. In addition some preliminary ESR results suggest that as in silicon the single vacancy migrates with a higher activation energy in *p*-type than in *n*-type germanium. In conclusion one might anticipate a more extensive use of the ESR technique applied to irradiation effects in germanium. The experience gained in Si may then be invaluable in unravelling the defect structure. However, even in Si much remains to be done and a more reliable theoretical method of investigating the defects in Group IV elements needs to be developed.

5.4 Dislocations in semiconductors

It is comparatively easy to study the structure of dislocations in the semiconductors, since single crystals of these materials may in general be grown to high purity. Furthermore the crystals cleave relatively easily. Both silicon and germanium cleave on the {111} planes, and in this respect they differ from the compound semiconductors with the zinc-blende structure, where the {110} planes are the preferred cleavage planes. This may be due to the ionic contribution to bonding in the III to V compounds. In the zinc-blende structure the {111} planes are formed by layers of identical atoms. Thus two successive layers of different atoms will attract more strongly than two layers of {110} planes, which are built up equally out of the two different kinds of atoms.

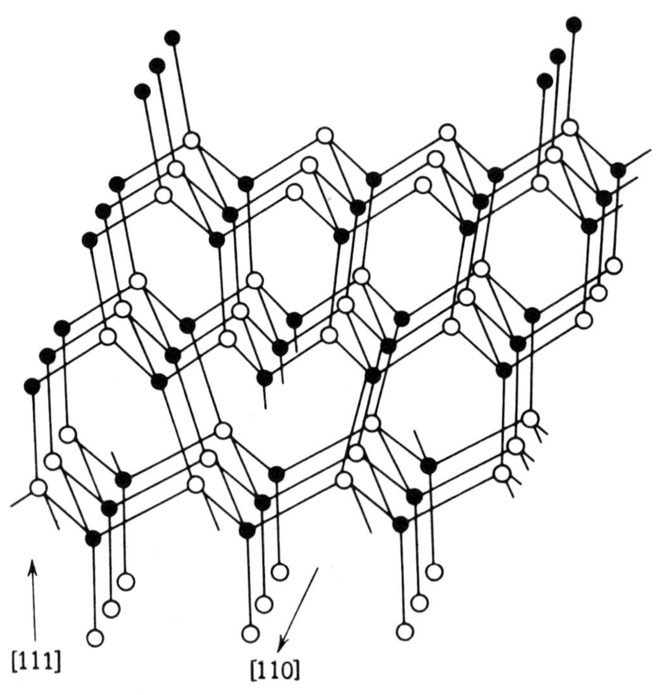

Fig. 5.14 The structure of the "60° dislocations" in the diamond (all atoms identical) and zinc-blende structures. (After MADELUNG. O., 1964, Physics of III. V compounds Wiley, New York.)

The dislocation structures have been studied in numerous ways. Etch pit techniques and impurity decoration techniques were much used in the early investigations. More recently transmission electron microscope techniques have been successful.[50] However, in this section we draw attention only to an essential difference between the dislocation structures in the elements and compounds. This can be seen by reference to Fig. 5.14 which represents a "60° dislocation" parallel to the ⟨110⟩ direction. It can be seen that along the core of this dislocation there are a number of atoms which are bonded to only three neighbouring atoms. In germanium and silicon these atoms have one free valence electron. Consequently they can act either as acceptors by trapping a hole or as donors by taking on a second electron to form a saturated electron pair bond. The "60° dislocation" in compound semiconductors will of course show different behaviour according to whether the row of unsaturated bonds is associated with Group III or Group V atoms. In the former case there is no excess of electrons and such a row of atoms will act as an acceptor level. When Group V atoms form this row, each atom has a saturated electron pair and the dislocation introduces a donor level.

It will be recognized that these electrically active dislocations will have important effects on numerous properties. In indium antimonide the electron mobilities are decreased by dislocations if the electrons move perpendicular to the bending axis of deformation. For current parallel to the axis little change in the mobility is detected. This fall in mobility is interpreted as due to additional scattering by dislocations lying parallel to the bending axis. In view of their electrical character these electrically active dislocations may also have an important bearing on the diffusion of differently charged impurities through the lattice of compound semiconductors.

GENERAL REFERENCES

BLAKEMORE, J. S. 1969. *Solid State Physics.* Saunders, Philadelphia.
MADELUNG, O. 1964. *Physics of III–V Compounds.* Wiley, New York.
SEEGER, A. and CHIK, K. P. 1968. *Phys. Stat. Solid.* **29**, 455.
CORBETT, J. W. 1966. *Electron Irradiation in Metals and Semiconductors.* Academic Press, New York.

6

Point Defects in Metals and Alloys

6.1 Lattice vacancies at thermal equilibrium

In Chapters 3 to 5 we discussed the properties of point defects in insulators and semiconductors. The proposed defect structures were determined usually by spectroscopic methods. Such methods are not easily applied to point defects in metals. Instead indirect techniques such as electrical conductivity, lattice parameter and crystal density measurements have found wide application. Under special circumstances microscopic techniques are possible.

Lattice parameter and crystal density studies

Whether Schottky defects or Frenkel defects are predominant in a solid at thermal equilibrium may be determined by comparative measurements of unit cell size a and macroscopic crystal dimensions L. In most metals at temperatures up to melting $\Delta L/L - \Delta a/a > 0$, indicating the presence of Schottky defects in thermal equilibrium (see §2.3). Since $\Delta a/a$ is affected only by the dilatation around defects whilst $\Delta L/L$ responds to the overall dimensions of defects including the dilatations, Equ. 2.2 becomes

$$n_s = 3\left(\frac{\Delta L}{L} - \frac{\Delta a}{a}\right) \tag{6.1}$$

Equation 6.1 is independent of the dilatations around defects because such dilatations affect $\Delta L/L$ and $\Delta a/a$ equally. Consequently simultaneous measurements of the unit cell and linear macroscopic expansions as a function of temperature should yield equilibrium point defect concentrations, n_s, with high precision. This comparative method has been widely used in face-centred cubic metals. A comparison of $\Delta L/L$ and $\Delta a/a$ for aluminium in the temperature range 229° to 655°C is shown in Fig. 6.1a. Only above 420°C is $\Delta L/L$ noticeably greater than $\Delta a/a$. The net fractional changes in the single vacancy concentration as a function of temperature are portrayed in Fig. 6.1b. Assuming only single vacancies to be present these results fit

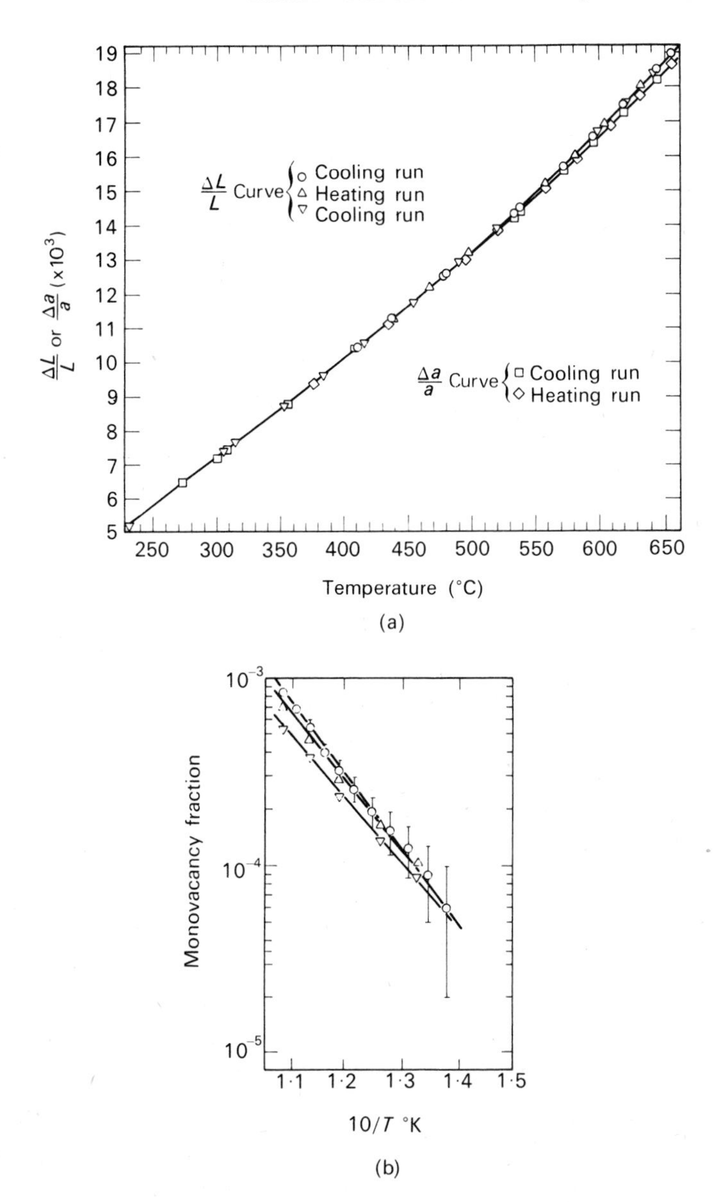

Fig. 6.1 Showing the temperature dependence of (a) $\Delta L/L$ and $\Delta a/a$ of aluminium and (b) the monovacancy concentration in Al derived from (a) assuming $E_b^2 = 0$ (○), $E_b^2 = 0{\cdot}25$ eV (△) and $E_b^2 = 0{\cdot}35$ eV (▽). (After SIMMONS and BALLUFFI. 1960. Phys. Rev. **117**, 52.)

the equation,

$$n_S = \exp(S/k) \exp(-E_S/kT) \tag{6.2}$$

with the vacancy formation entropy $S/k = 2{\cdot}2$ and the vacancy formation energy $E_S = 0{\cdot}76$ eV.

The possibility that divacancies, trivacancies, etc., are present at elevated temperatures must not be overlooked. Defect aggregates have concentrations given by,

$$n_x = a_x \exp(-E_x/kT)$$

x being the number of vacancies in the aggregate, a_x is a geometrical factor related to the number of ways of arranging the multiple vacancy in the lattice and E_x is the free energy of formation of the multiple vacancy. Clusters of vacancies in nearest neighbour sites are most probable since their formation energy is small relative to less close-packed clusters on account of their increased binding energy. Including divacancies the total vacancy concentration is

$$n_s = n_1 + 2n_2$$

since each divacancy involves two single vacancies. Hence

$$n_s = \exp - (E_S/kT) + 12 \exp(-E_{S2}/kT) \tag{6.3}$$

The free energy of divacancy formation E_2 is related to the binding energy E_2^b by

$$E_2^b = 2E_S - E_{S2} \tag{6.4}$$

which on substitution in Equ. 6.3 gives

$$n_s = n_1 + 12n_1^2 \exp(E_2^b/kT) \tag{6.5}$$

In Fig. 6.1b are plotted the single vacancy concentrations for assumed divacancy binding energies $E_2^b =$ OeV, 0·25 eV and 0·35 eV. It appears probable that E_2^b is closer to the value of 0·15 eV.[51] At the melting point of aluminium the total vacancy concentration n_s is $9{\cdot}4 \times 10^{-4}$, this result being independent of the detailed structure of the defects. The divacancy contribution to the defect population is then less than 10% for an assumed binding energy of 0·15 eV.

Similar measurements have also been made on other face-centred cubic metals and representative results are given in Table 6.1. It is evident that n_s is much higher for gold than for silver and copper, so that the formation entropy S_f and the formation energy E_f can be determined directly, as was the case for aluminium. The value of S for Ag is obtained by comparison of the

Table 6.1 Some properties of monovacancies in face-centred cubic metals

Metal	n_s at M. Pt. ($\times 10^4$)	S_f/k	E_f	E_M	$\rho/1\%$ vacancies ($\mu\Omega$. cm)
Cu	2·0 ± 0·5	—	1·17 ± 0·11	0·88 ± 0·05	—
Ag	1·7 ± 0·5	1·5 ± 0·5	1·09 ± 0·05	0·83	1·3
Au	7·2 ± 0·6	1·2 ± 0·2	0·96 ± 0·4	0·82	1·5
Al	9·0 ± 0·6	2·2	0·76 ± 0·7	0·65	3·0
Pb	—	1·5	⩾0·53	—	
Pt	—	—	1·2	1·4	

direct equilibrium measurements with quenching studies. The measured formation entropy S lies between one and two entropy units, which is to be expected from the structural form of the monovacancies. These results are all consistent with 90% of the vacant lattice sites being monovacancies. Equilibrium vacancy concentrations have not been measured with the same degree of confidence in alloys.

Resistivity measurements

The measurement of electrical resistivity is a simple and sensitive means of studying structural defects in metals. Simple theory divides the resistivity into two contributions, one due to electron scattering by the "perfect" lattice and one due to lattice imperfections (§2.4). The total resistivity given by Equ. 2.6 may be expanded as

$$\rho(T) = \rho_0 + \rho_i(T) + C \exp - (E_s/kT) \tag{6.2}$$

The residual resistivity ρ_0 represents the temperature-independent contribution from static lattice defects, whilst $\rho_i(T)$ is the temperature dependent contribution due to scattering by lattice vibrations. Finally the term $C \exp(-E_s/kT)$ represents a contribution due to the equilibrium concentration of lattice defects.

No fully justifiable theoretical method exists to extrapolate the ideal lattice resistivity to high temperature, even for simple metals. For aluminium Simmons and Balluffi[51] extrapolated the resistivity below 320°C to high temperature using two different expansions viz

$$T = A_0 + A_1 \ln (\rho_i(T)/T) + A_1 \ln (\rho_i(T)/T^2)$$

and

$$\rho_i(T) = B_0 + B_1 T + B_2 T^2 \tag{6.6}$$

with seemingly little difference in results for the formation energy. At high temperature the defect-induced resistivity increase is then obtained by taking differences between the resistivity estimated using Equ. 6.6 and the measured resistivity. The data for Al are well fitted by the relation

$$\Delta\rho = (4{\cdot}4 \times 10^{-3}\,\Omega\ \text{cm}) \exp(-0{\cdot}77\ \text{eV}/kT)$$

This value of $E_s = 0{\cdot}77$ eV is in good agreement with the value determined by measurements of dimensional changes. At the melting point the resistivity increase due to thermally induced defects amounts to 3 $\mu\Omega$ per 1 atomic percent of vacancies. In view of the uncertainties in the extrapolation procedure, the method is not widely used.

Nonstoichiometry in intermetallic phases

Excess lattice vacancies are sometimes formed when intermetallic phases such as the γ-brasses, transition metal aluminides and transition metal carbides deviate from the stoichiometric composition. The complex cubic unit cell of the γ-brasses contains 52 atoms when all atomic sites are occupied. At stoichiometry the alloy composition corresponds to a valence electron/atom ratio of 21/13 in systems such as Cu–Zn, Ni–Zn, Cu–Ga and Cu–Al. A survey of γ-phases in various alloy systems shows that although their compositions vary on either side of stoichiometry, an electron concentration corresponding to 88 electrons per unit cell is a critical limit beyond which the normal γ-structure cannot exist.* Detailed comparisons of the lattice parameter and density of the γ-phases Cu_9Al_4 and Cu_9Ga_4 as a function of composition shows that the number of atoms in the unit cell is not necessarily constant. For compositions corresponding to electron/atom ratios greater than 1·7 the number of atoms in the unit cell actually falls. The extent to which atoms are omitted from the structure corresponds to the maintenance of a constant number of 88 electrons per unit cell. If the concentration of solute atoms is increased above this limit then either atoms are omitted to preserve this constant electron concentration or a new phase is formed.

Similar behaviour is observed in the transition metal aluminides. These compounds have the ordered body-centred cubic β-brass structure and are stable at compositions corresponding to electron/atom ratios near 3/2. As in the γ-brasses the vacant lattice sites are formed in such a way that the number of electrons per unit cell remains constant. NiAl is a typical β-brass compound, stable over a wide range of homogeneity about the 50/50 composition. The influence of the vacant lattice sites on the lattice parameter and density of NiAl is illustrated in Fig. 6.2. Since Ni atoms are smaller than Al, an increase in the Ni content above 50% causes the lattice parameter to

* For a critical review of the occurrence and stability of inter-metallic compounds see HUME-ROTHERY and RAYNOR, 1956. *The Structure of Metals and Alloys*. Inst. of Met., London.

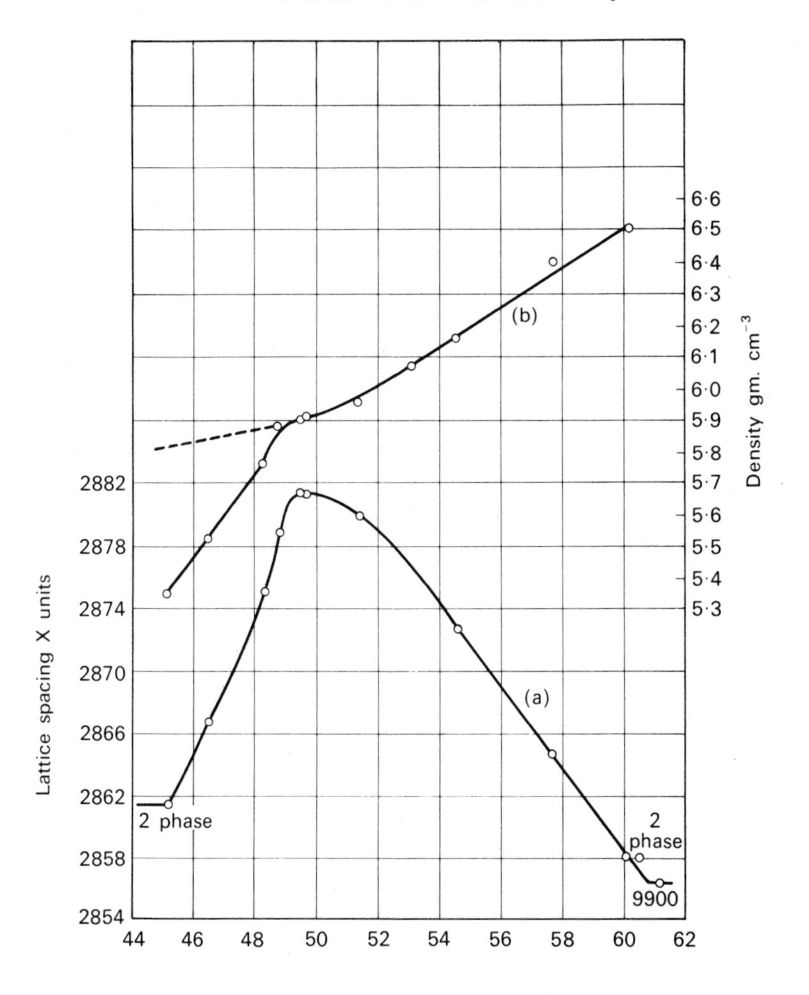

Fig. 6.2 Showing the lattice parameter and density variations of the β-brass compound NiAl as a function of composition. (After BRADLEY and TAYLOR. 1937. *Proc. Roy. Soc.* **159,** 56.)

decrease and the density to increase. Thus the decrease in both lattice parameter and density when the Ni content is less than 50% is anomalous and occurs because some of the lattice sites vacated by Ni atoms are not occupied by Al atoms. Consequently as the composition varies towards pure Al, the compound takes up sufficient vacancies to maintain the number of electrons per unit cell at a constant value of 3. The effect is such that there are about 7% vacant sites on the Ni sublattice in an alloy containing 54 at. % Al.

The disposition of the vacant lattice sites has important consequences in respect of other studies. For example, when an alloy containing, say, 54 at. % Al is quenched from a higher temperature, it is expected that the vacancies will remain essentially isolated. In a slowly cooled alloy the vacancies form aggregates, although such an effect is not expected to result in marked density changes. According to Ball[52] the vacancy clusters are planar defects in which, as Fig. 6.3 shows, the excess aluminium atoms are accom-

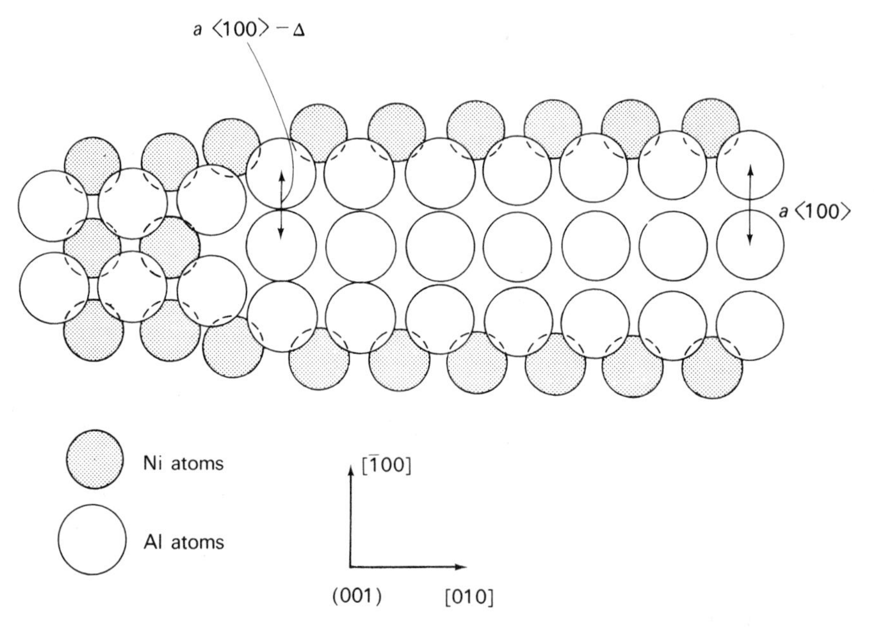

Fig. 6.3 A possible vacancy platelet configuration on the (100) plane of NiAl. This structure is consistent with an excess of aluminium atoms in non-stoichiometric NiAl. (After BALL. 1969. *Phil. Mag.* **20**, 113.)

modated as discs of atoms on normal lattice sites sandwiched between layers of nickel vacancies. Thus there are two vacant nickel sites created for each excess aluminium atom, in accord with the results in Fig. 6.2. Such stacking faults give rise to the easily recognizable contrast effects in Fig. 6.4a. It is evident both from the structure of the fault and the electron micrograph that such planar defects are terminated by dislocation lines.

The cubic transition metal carbides also show large deviations from stoichiometry as a result of carbon vacancies being incorporated in the face-centred cubic carbon sublattice. In titanium carbide the vacancies are randomly distributed in the lattice over most of the composition range. In vanadium carbide, however, the carbon vacancies take up ordered positions in the lattice. The extra lines which are then found in the X-ray diffraction

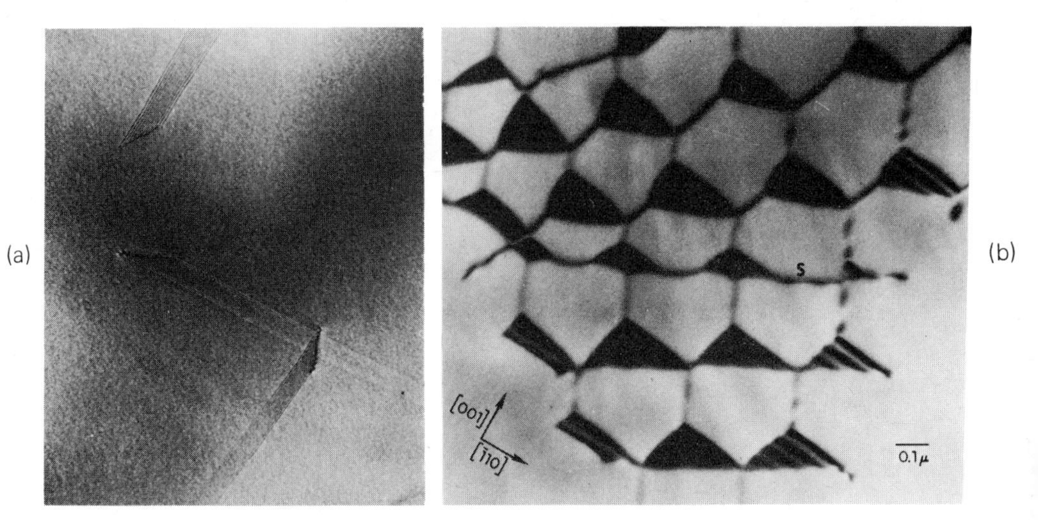

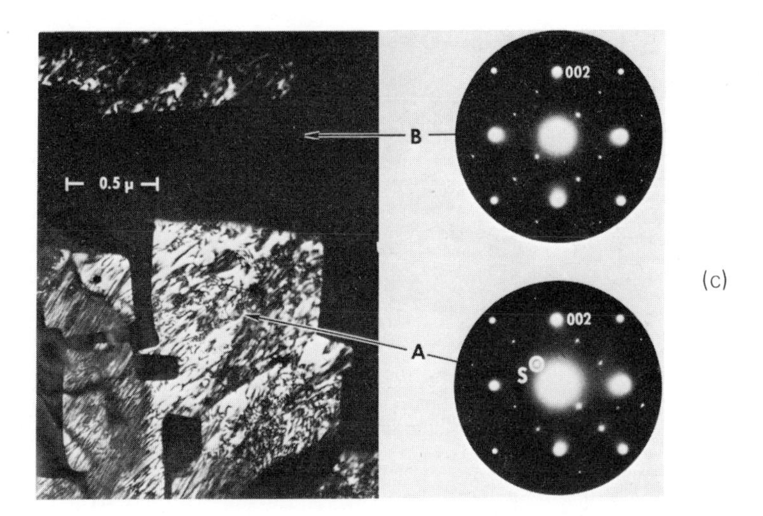

Fig. 6.4 Showing some consequences of non-stoichiometry on the microstructure of the compounds NiAl and VC. (a) Showing an electron micrograph of stacking faults on the cube planes of NiAl containing 53·8 atomic % Al. (After BALL[52].) (b) Optical micrograph of domain structure in V_6C_5 taken with polarized light between crossed nicols on (100) cleavage plane. (c) Dark field transmission electron micrograph of domain structure in $VC_{0{\cdot}84}$ taken with superlattice reflection S. Selected area diffraction patterns obtained from domains A and B demonstrate that the primary VC pattern maintains equivalent orientation in two regions, but superlattice pattern undergoes a 90° rotation. (After VENABLES *et al.*[53].)

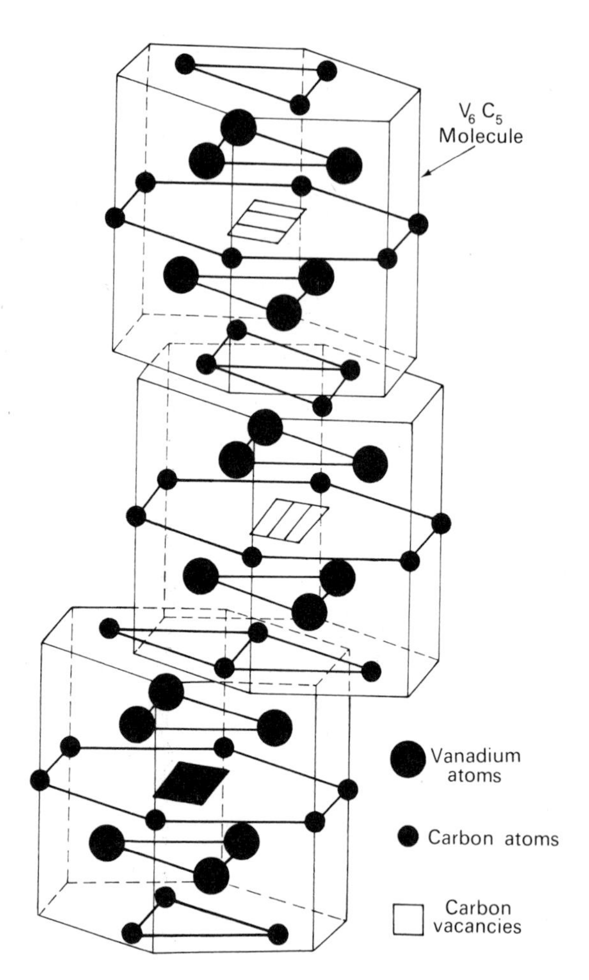

Fig. 6.5 Showing the nature of the ordered vacancy structure in non-stoichiometric VC. In particular the unit cell of the composition V_6C_5 is shown. (After VENABLES et al.[53])

pattern at the chemical composition V_8C_7 are attributed to a cubic superlattice resulting from vacancy ordering. The ordered structure of V_8C_7 has a lattice parameter twice that of the stoichiometric rock-salt unit cell. This disorder-order transformation corresponds to the unit cell changing from the symmetry group $F\frac{4}{m}\bar{3}\frac{2}{m}$ (rocksalt) to the symmetry group $P4_332$.*

* We use here the Hermann-Mauguin representation, according to which the structure is characterized by tetrads along the ⟨100⟩ axes, triads along the ⟨111⟩ axes and the diads along the ⟨110⟩ axes.[40]

The nature of the ordering in VC is strongly dependent on composition. In the compound V_6C_5 Venables *et al.*[53] have demonstrated the existence of the hexagonal superlattice shown in Fig. 6.5. That the structure is not cubic is shown by polarized light microscopy since a domain structure is observed in single crystals. The optical anisotropy causes different colourations as a result of birefringence, which are visible as black and white domains in Fig. 6.4b. Figure 6.4c shows an electron micrograph and the accompanying electron diffraction patterns of two adjoining domains. The main spots in the diffraction patterns maintain equivalent orientations in the different regions of the thin foil. However, the superlattice pattern is rotated through 90° between the two different domains. Thus the domain structure results because the superlattice can assume several possible orientations within the nominally cubic vanadium carbide lattice. A complete analysis of the superlattice diffraction patterns reveals that the ordered structure is hexagonal about one of the four equivalent $\langle 111 \rangle$ directions of the cubic lattice.

The diffraction pattern alone is not sufficient to determine the vacancy distribution within the unit cell. Nuclear magnetic resonance selects the 100% abundant ^{51}V nuclides ($I = 7/2$) for detailed examination. The nuclear electric quadrupole moment of this isotope is sensitive to the presence of an electric field gradient at the nucleus. Both the symmetry and intensity of the electric field experienced by ^{51}V are determined by the relative positions of the carbon vacancies. Thus the NMR line associated with ^{51}V nuclei will be split in a complex way depending upon the disposition of carbon vacancies around the ^{51}V nuclides. A typical spectrum from $V_6C_{5 \cdot 04}$ is shown in Fig. 6.6 in which it is possible to discern,

(i) weak components due to V atoms with no vacancies in the nearest shell of carbon atoms but with carbon vacancies in the more remote shells,

(ii) strong components due to $\frac{2}{3}$ of the V nuclei which have a vacancy in the nearest neighbour shell but also a vacancy in the second nearest neighbour shell of carbon atoms,

(iii) strong components due to $\frac{1}{3}$ of the V nuclei which have three neighbouring carbon vacancies, one in the nearest neighbour shell and two in the next nearest neighbour shell. The total intensity of all the lines correspond to a composition of $V_6C_{5 \cdot 028}$, in very close agreement with the known chemical composition.

The detailed model must conform with these NMR results and also disperse the vacancies in an ordered manner on the carbon sublattice so that the overall structure is hexagonal. A schematic representation of the structure in Fig. 6.5 shows that the trigonal lattice lies parallel to the $\langle 111 \rangle$ direction of the face-centred cubic titanium sublattice. Since this triad is also a screw axis, it can be seen that the carbon vacancies form a spiral around this axis, each vacancy being separated by a $\langle 112 \rangle$ distance from its neighbouring

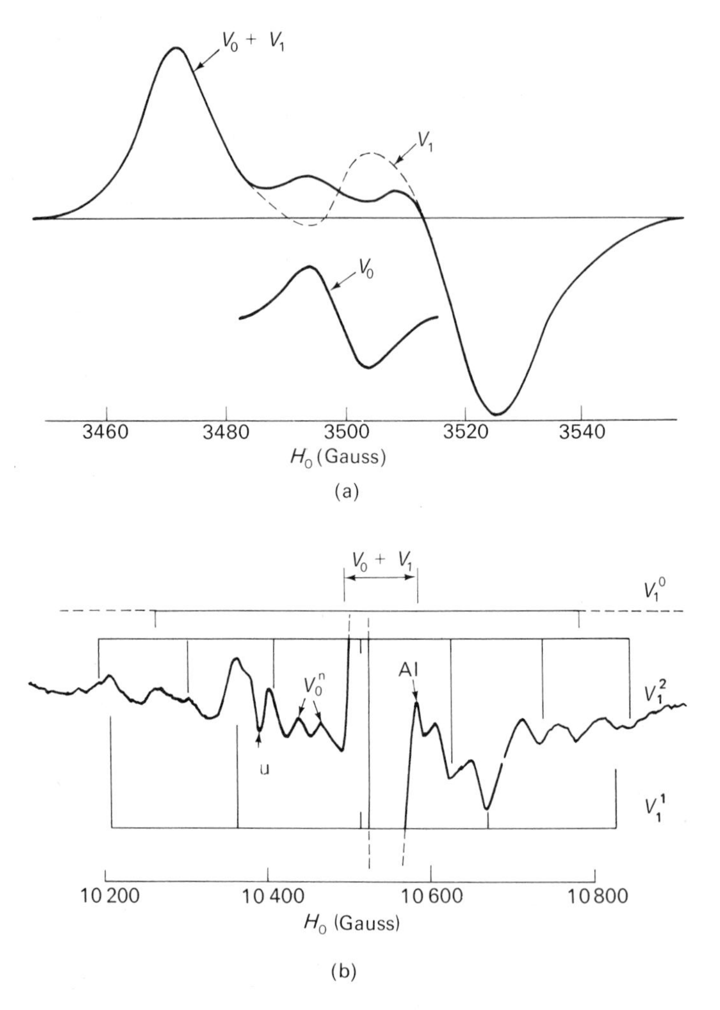

Fig. 6.6 NMR spectra at 77 K of ^{51}V nuclei in V_6C_5. The central resonance (a) is separated into two components due to nuclei with no nearby C vacancy (V_0) and with one nearby vacancy (V_1). Most of the satellite structure in (b) is attributed to resonances from nuclei at three different sites relative to nearly vacancies. (After VENABLES *et al.*[53].)

vacancies. The resulting structure belongs to either the $P3_1$ or $P3_2$ class. By comparison the V_8C_7 superlattice results when the vacancies form a spiral around the tetrad axis. It is apparent that the previous designations that the V_8C_7 and V_6C_5 compositions fall within the cubic NaCl-structured phase field, are inexact and should be amended to be consistent with these recently identified structures.[53]

6.2 The defect structure of quenched metals

Electrical resistivity measurements

In metals the vacancy formation energy is easily and accurately assessed by measurements of the electrical conductivity of specimens quenched from high temperature. This technique is advantageous since vacancy concentrations typical of high temperature equilibrium may be studied at low temperature, and the extrapolation procedures of §6.1 are unnecessary. The somewhat surprising scatter in the many experimental results for simple metals such as gold and silver has occurred because little account was taken in early work of such experimental variables as quenching rate and temperature, quenching strains and cluster formation. For gold these factors are minimized by quenching from temperatures below 600°C, where Kino and Koehler[54] observed that the quenched-in resistivity obeys the equation

$$\Delta\rho = (3{\cdot}6 \times 10^{-4}\,\Omega\,.\,\text{cm}) \exp(-0{\cdot}94\,\text{eV}/kT)$$

It follows that the activation energy so determined is in excellent agreement with values obtained from dimensional changes.

Vacancy migration energies may be determined from measurements of the annealing of the quenched-in resistivity. The most convenient means of doing this is by the 'change of slope method': ρ is measured as a function of time during isothermal annealing at a temperature T_1. The temperature is changed abruptly to a higher temperature T_2 and again the time dependence of ρ is measured. If n is the number of defects present at any time t, then

$$-(dn/dt) = An^x \exp(-E_m/kT)$$

represents the annealing characteristics for a reaction of order x, where E_m is the migration energy. In terms of the resistivity changes

$$\boxed{-\text{d}(\Delta\rho)/\text{d}t = A(\Delta\rho)^x \exp(-E_M/kT)} \qquad (6.7)$$

At the point on the annealing curve where T is suddenly increased from T_1 to T_2 $\Delta\rho$ is the same, hence

$$\boxed{E_M = k\left(\frac{1}{T_2} - \frac{1}{T_1}\right) \log_e \left[\frac{R(T_2)}{R(T_1)}\right]} \qquad (6.8)$$

where R refers to the rate of annealing at the particular temperature.

Some results for gold quenched from 691°C and subsequently annealed at 140°C and 150°C are shown in Fig. 6.7. Since according to Equ. 6.7 $\Delta\rho$ is linear in time for first order annealing kinetics, a higher order process must be taking place, of which the formation and migration of divacancies is the most likely. Second order kinetics then apply and it becomes possible to determine the activation energies for monovacancy and divacancy migration as well as the divacancy binding energy. The results of Kino and Koehler's detailed studies of gold may be summarized as follows[54]

Single vacancy $E_F = 0{\cdot}95 \pm 0{\cdot}05$

$E_M = 0{\cdot}94 \pm 0{\cdot}05$ at 150°C and low concentration

$E_M = 0{\cdot}85 \pm 0{\cdot}03$ at high temperature

Divacancy $E_b = 0{\cdot}15$ eV

$E_M = 0{\cdot}70 \pm 0{\cdot}10$ eV

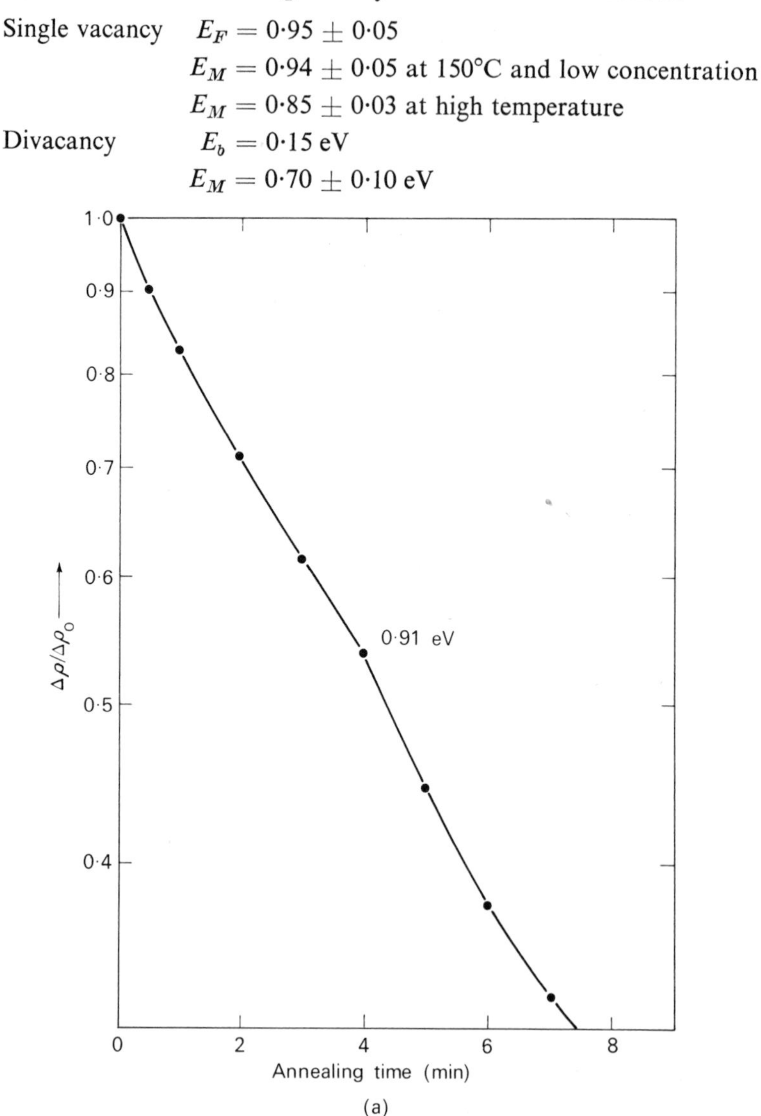

Fig. 6.7 Isothermal annealing curves for gold quenched from 691°C. After annealing at 140°C, the temperature was changed to 150°C. The activation energy for the recovery process may be calculated from equation 6.8 using the slopes of the curve in (a) where the temperature has been changed. (KINO and KOEHLER. 1967. *Phys. Rev.* **162,** 632.)

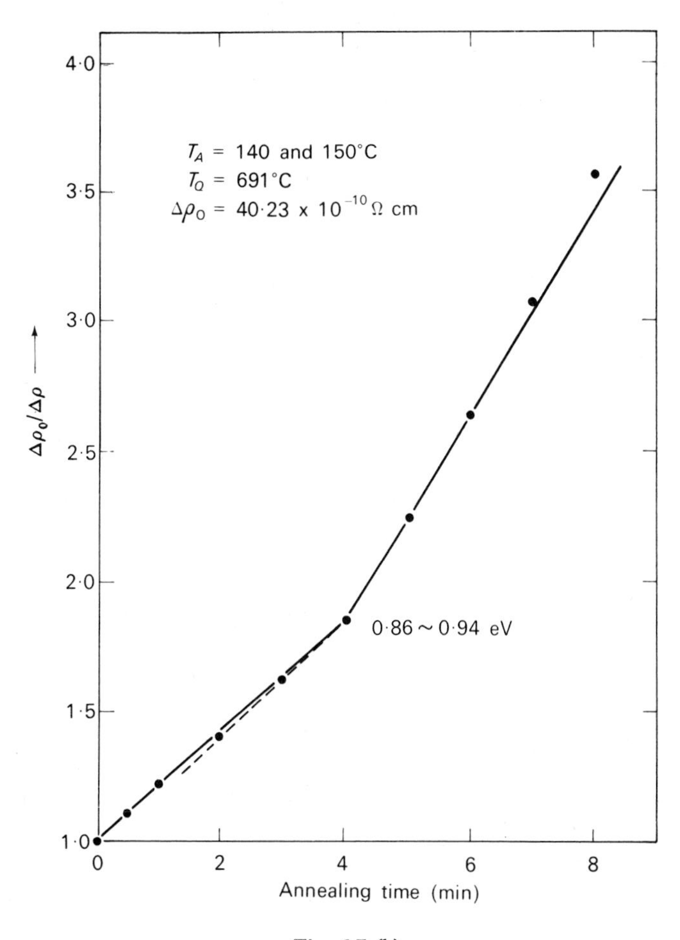

Fig. 6.7 (b)

The temperature dependence of the migration energy is reasonable since the lattice is more open at high temperature and hence less energy is required to overcome the cohesive forces in the lattice. We emphasize that the measurements described above were mainly done on gold specimens quenched from below 600–700°C. In this way, vacancy losses to other sinks as well as the formation of vacancy pairs and higher order aggregates are minimized.

Electron microscopy of quenched metals

It is clear that quenching from low temperatures (say $T_Q < \frac{2}{3}T_m$) is an effective means of retaining a super-saturation of *isolated* vacancies in metals. At higher temperatures where the vacancy mobility is high, migration to a variety of available vacancy 'sinks' (e.g. dislocations, grain boundaries etc.) is probable. When vacancies migrate to edge dislocations the dislocation lines climb normal to their slip plane. This results in the formation of *jogs*. Obviously since large numbers of vacancies do not arrive at the dislocation line uniformly the climb process is discontinuous, occurring one atom at a time. Thus the edge dislocation will be composed of numerous sections of different lengths, lying on neighbouring parallel slip planes and joined

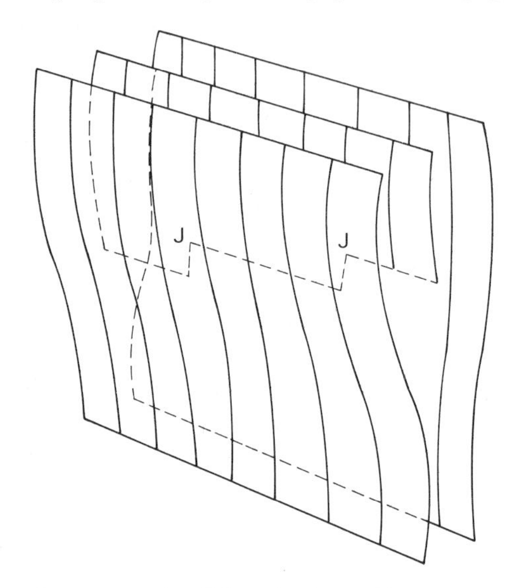

Fig. 6.8 Jogs on an edge dislocation formed by vacancy migration.

together by jogs (Fig. 6.8). Since dislocation climb requires mass transport by diffusion it is thermally activated.

Screw dislocations also climb as a result of vacancy condensation; in so doing they result in the long spirals or helices shown in Fig. 6.9a. The large individual dislocation loops also shown in Fig. 6.9a have almost certainly formed as a result of the breakdown of the dislocation spirals. This electron micrograph was obtained by quenching an Al-6% Mg alloy from 550°C into distilled water followed by annealing at 100°C for 20 minutes. Dislocation helices are not observed in pure aluminium because there are no impurity atoms available to lock the helical dislocation along its path. Thus helical dislocations straighten out in pure metals as a result of the inherent line tension.

Dislocation loops also form in aluminium as a result of vacancy condensation onto the {111} planes, as we mentioned in §1.2. In aluminium foils quenched from near the melting point the loops are frequently observed to be 200 to 3 000 Å in diameter. However the distribution of loops throughout the specimen is by no means uniform. Near a grain boundary there is a

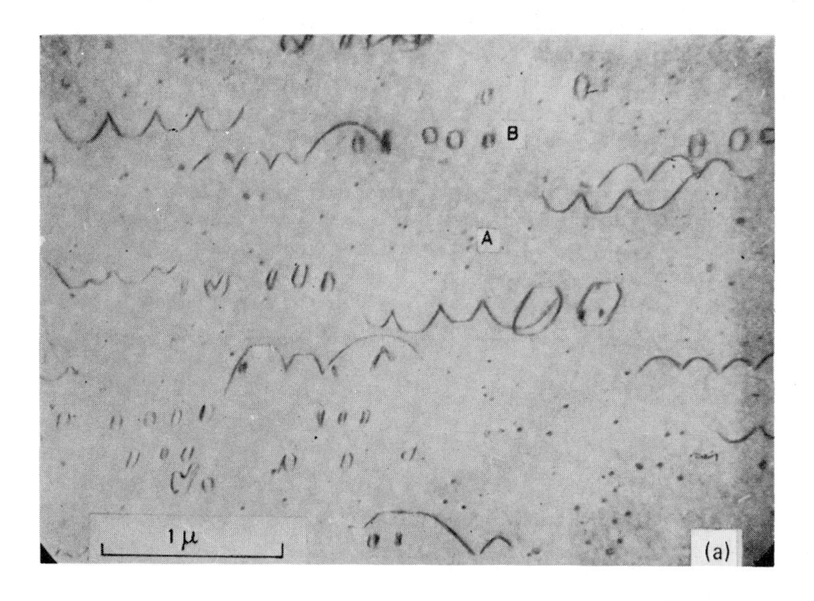

Fig. 6.9 (a) Helical dislocations and prismatic loops in Al-6% Mg alloy quenched from 550°C into water and annealed at 100°C for 20 minutes. (reproduced by kind permission of Prof. R.E. Smallman.)

zone, extending on either side of the boundary, which is depleted of dislocation loops (Fig. 6.9b). Similar effects are observed also near dislocation lines. Such depletion zones are not unexpected: both grain boundaries and dislocation lines act as sinks for vacancies so that the degree of supersaturation near them is always too low to cause loop formation.

Figure 1.4a shows that when a platelet or disc of vacancies condenses on a close-packed plane a stacking fault ringed by a dislocation line is produced. The electron microscope images of these defects should be characterized by diffraction fringes, bands of light and dark contrast, inside the dislocation loop. Since Fig. 1.5a shows that in the majority of cases no such stacking fault contrast is observed, and most of the quenched-in defects must be perfect dislocations. Consequently, the fault shown in Fig. 1.4a cannot be the most stable structure. The reason for this can be seen in the following way. When a partial layer of atoms is removed as in Fig. 1.4a, the stacking sequence across the fault becomes ABC AC ABC. The boundary between this fault and the perfect crystal is an edge dislocation with Burgers vector $(a/3)$ $\langle 111 \rangle$ perpendicular to the fault plane. Since this Burgers vector is not contained in a close-packed plane the dislocation cannot glide conservatively under an applied stress. This is a *sessile* dislocation, or more particularly a *Frank sessile* dislocation.

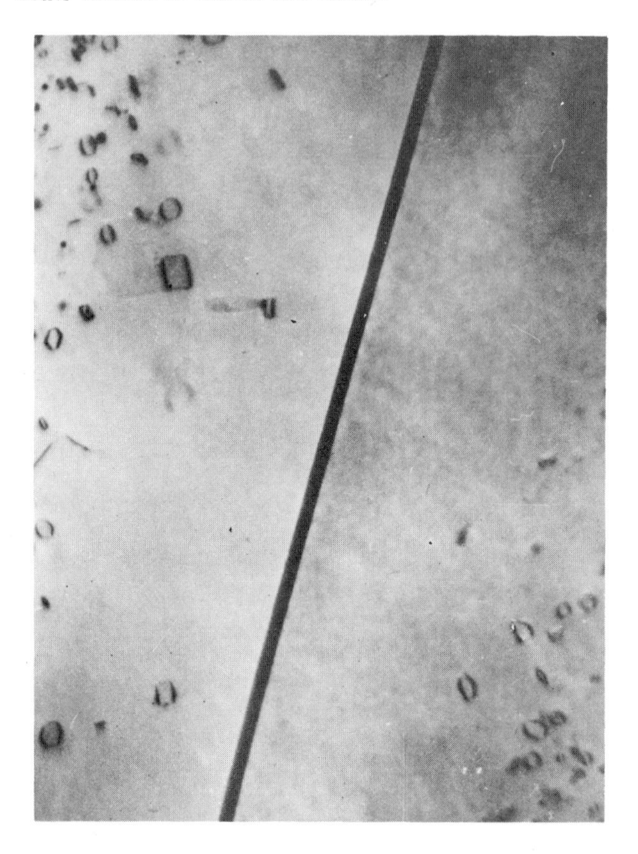

Fig. 6.9 (b) A zone denuded of loops in the region of a grain boundary in quenched aluminium. (After HIRSCH, SILCOX, SMALLMAN and WESTMACOTT. 1958. *Phil. Mag.* **3**, 897.)

When the stacking fault energy is high Frank sessile dislocations and their associated stacking faults are unstable. Under such circumstances it may be energetically favourable to nucleate a partial dislocation in the plane of the fault which then sweeps through the fault. The stacking fault in Fig. 1.4a is then removed provided that the lattice above the fault is sheared by one of the three $(a/6)\langle 112\rangle$ partial dislocations such that $C \rightarrow B$, $A \rightarrow C$, $B \rightarrow A$ etc. At the perimeter of the fault the partial dislocation interacts with the Frank sessile dislocation according to $(a/3)[11\bar{1}] + (a/6)[112] = (a/2)[110]$. Thus the dislocation loop in this case is a unit dislocation with Burgers vector $(a/2)[110]$: this is a prismatic loop which may glide along the surface of a cylinder in the [110] direction. That both Frank sessile loops and prismatic loops are observed in quenched aluminium suggests that the detailed energy balance is only just in favour of fault annihilation.

Neglecting external influences and thermal fluctuations the only force tending to nucleate the dislocation required to annihilate the fault is that due to the stacking fault energy γ. According to Cottrell[2] dislocations are nucleated in a perfect crystal when the shear stress is $\sim\mu/30$. Thus the fault, which produces a force of γ dynes, will be removed spontaneously if,

$$\gamma = \mu b/30$$

b being the Burgers vector of the partial dislocation. Assuming values of $\mu = 3 \times 10^{11}$ dyn. cm^{-2} and $b = 1{\cdot}5 \times 10^{-8}$ cm, the minimum stacking fault energy necessary for fault removal is $\sim$·000015 Joules cm^{-2}. In metals with stacking fault energies greater than this, dislocation loops will be unfaulted and consist of unit dislocations. It is inferred from the electron micrographs that the stacking fault energy of aluminium must be close to ·000015 Joules cm^{-2}.

Actually, these loops may be used to determine the stacking fault energy with high precision. The experimental method involves studying the kinetics of loop annealing. It is particularly applicable to materials with high stacking fault energy and potentially useful over a wide range of fault energies. On annealing thin aluminium foils the prismatic loops and Frank sessile loops shrink. The shrinkage process is controlled by the diffusion of vacancies away from the loop. The annealing rate of a dislocation loop depends upon the vacancy concentration gradient developed between the loop and the foil surface. The rate equations take the form,[55]

$$\boxed{\left(\frac{\mathrm{d}r}{\mathrm{d}t}\right)_F = -\frac{2\pi D}{b \ln (L/b)}\left[\exp\left(\frac{\gamma B^2}{kT}\right) - 1\right]} \tag{6.9}$$

for faulted loops and,

$$\boxed{\left(\frac{\mathrm{d}(r^2)}{\mathrm{d}t}\right)_P = -\frac{2\pi\alpha b D}{b \ln (L/b)}} \tag{6.10}$$

for prismatic loops, where γ is the stacking fault energy, D the self-diffusion coefficient, L the foil half-thickness and B^2 the cross sectional area of the vacancy on the (111) plane. By eliminating D between Equ. 6.9 and 6.10, the self-diffusion coefficient is removed as a possible source of error, and accurate values of γ may be subsequently determined. Thus we obtain

$$\boxed{\alpha b\left(\frac{\mathrm{d}r}{\mathrm{d}t}\right)_F \Big/ \left[\frac{\mathrm{d}(r^2)}{\mathrm{d}t}\right]_P = \exp(\gamma B^2/kT) - 1} \tag{6.11}$$

Usually the hexagonal-shaped loops become circular almost immediately on annealing, and remain uniformly circular during subsequent anneals. A typical sequence of electron micrographs depicting the climb of faulted

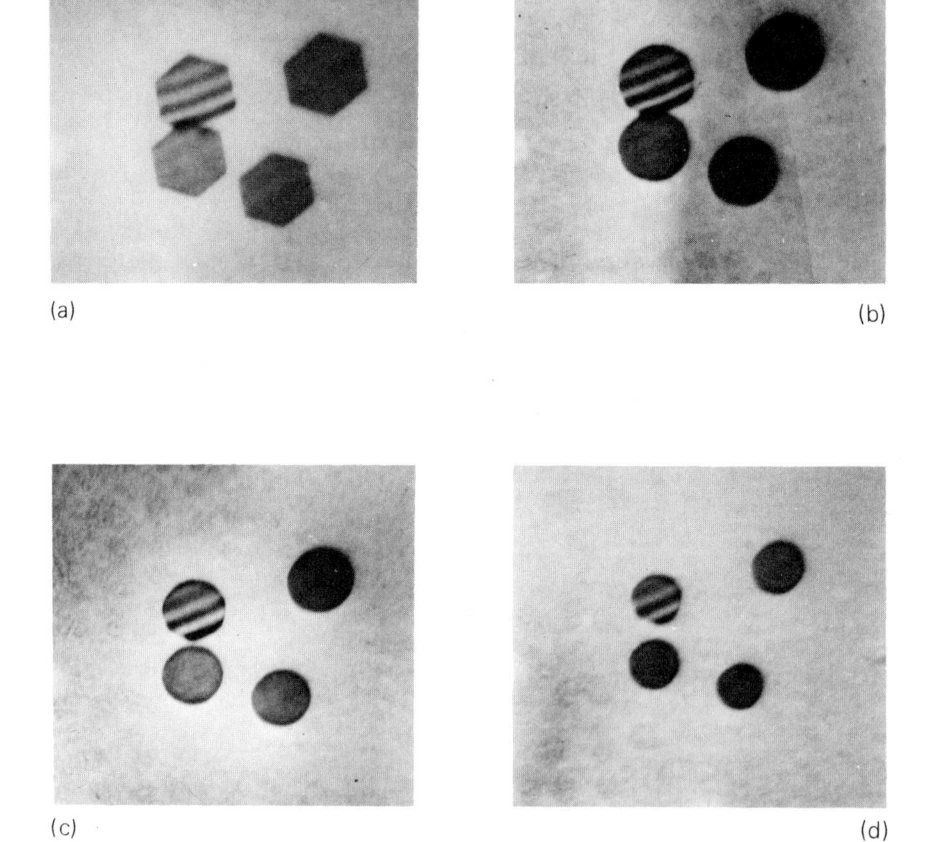

Fig. 6.10 Showing the climb of faulted dislocation loops in aluminium at 140°C. Annealing time (a) t = 3 mins (b) t = 6 mins, (c) t = 18 mins, (d) t = 30 mins. (After DOBSON, GOODHEW and SMALLMAN. 1967. *Phil. Mag*. **16**, 9.)

loops in aluminium as a function of time at 140°C is shown in Fig. 6.10. Dobson, Goodhew and Smallman[56] were not able to measure annealing rates on faulted and unfaulted loops at the same temperature. Consequently faulted loops were observed at temperatures between 130° and 150°C and prismatic loops between 170° and 200°C. The results were then interpolated as shown in Fig. 6.11 to give annealing rates for faulted and unfaulted loops

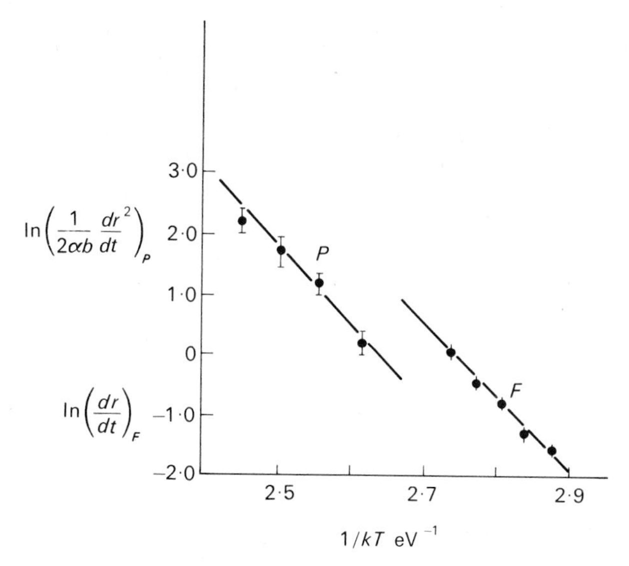

Fig. 6.11 Representing the annealing kinetics of faulted loops (F) and prismatic loops (P) in aluminium. The change in loop radius with time at temperature is plotted as the logarithm of (i) $(2\alpha b)^{-1}d(r^2)/dt$ for prismatic loops (line P) and dr/dt for faulted loops (line F) versus reciprocal temperature. (After DOBSON, GOODHEW and SMALLMAN. 1967. *Phil. Mag.* **16,** 9.)

at one temperature, which were then used in Equ. 6.11. The stacking fault energy of aluminium determined in this way is $13{\cdot}5 \pm 2\mu J \cdot cm^{-2}$. The method has been applied also to the climb kinetics of dislocation loops in magnesium and zinc.

Gold has a relatively low stacking fault energy and vacancy clustering should not result in prismatic dislocation loops being formed. However, electron microscope examinations of pure gold foils quenched from near melting point reveal defects which exhibit contrast effects typical of stacking faults (Fig. 6.12). These defects are not sessile dislocation loops lying on a (111) plane, but are tetrahedral defects made up of stacking faults on all four (111) planes. They are believed to form by vacancy condensation on a single (111) plane, nucleating a stacking fault with edges parallel to three ⟨110⟩ directions and bounded by Frank sessile dislocations. This configuration achieves a lower energy by sessile dislocation dissociation according to

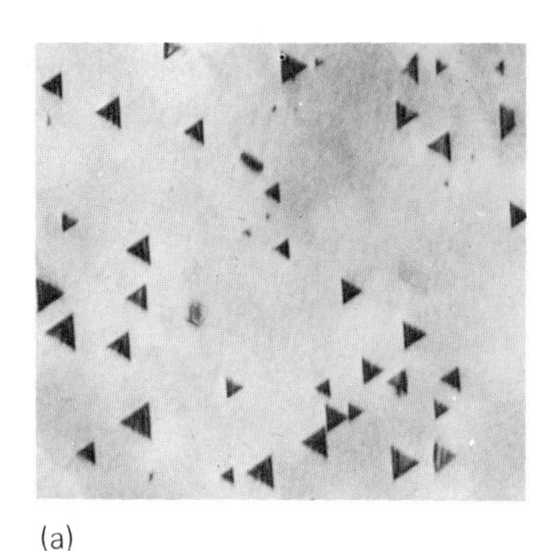
(a)

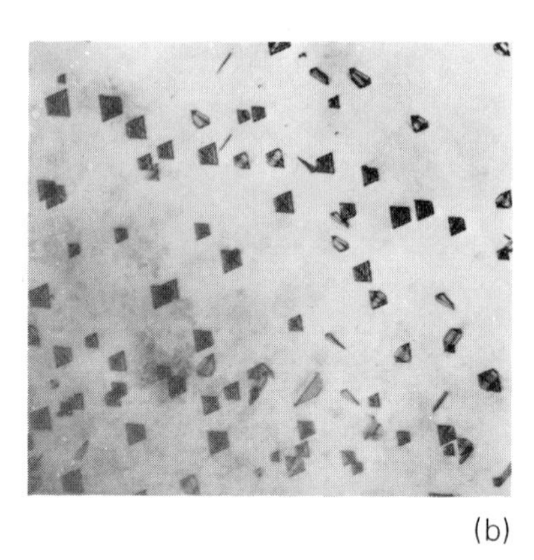
(b)

Fig. 6.12 Transmission electron micrograph of tetrahedral loops in quenched gold in the (a) (110) foil orientation and (b) (100) foil orientation. (reproduced by kind permission of Prof. R. E. Smallman.)

$(a/3)\ [111] \rightarrow (a/6)\ [121] + (a/6)\ [101]$. The partial dislocations $(a/6)\ [121]$ being repelled by the $(a/6)\ [101]$ dislocations, spread out on the three neighbouring $\{111\}$ planes trailing stacking faults in their wake. They are then attracted together in pairs to form new dislocations $(a/6)\ [\bar{1}10]$ thus:

$$a/6\ [121] + a/6\ [\bar{2}\bar{1}\bar{1}] = a/6\ [\bar{1}10]$$

Since the dislocation energy is proportional to b^2 the total energy is reduced and the reaction energetically favoured. Consequently when vacancies condense on a (111) plane to form a Frank sessile dislocation with segments lying along the three $\langle 110 \rangle$ directions, dissociation reactions result in the creation of stacking faults on the other three planes. Such tetrahedral imperfections would appear triangular when viewed along a $\langle 110 \rangle$ direction as shown in Fig. 6.12a and square when viewed along a $\langle 100 \rangle$ direction (Fig. 6.12b).

To emphasize the role played by the stacking fault energy as a microstructure-determining parameter we note that in copper which has a stacking fault energy appreciably larger than that of gold, both prismatic loops and spiral dislocations are formed. Pure silver which, has a stacking fault energy rather smaller than that of gold, shows prismatic loops and stacking fault tetrahedra.

6.3 Radiation damage in metals

The defect structure of irradiated metals varies with the type of irradiation, heavy particles producing the most extensive damage. In the case of electron irradiation the damage is likely to consist of interstitial-vacancy pairs randomly distributed throughout the crystal volume. The separation between defect pairs depends upon the displacement energy of the material and upon the radiation energy and dose. Electron bombardment has two unique advantages in studying irradiation damage both of which stem from the small energy transfer between the electrons and the lattice atoms. Since the recoil energy is small, isolated Frenkel defects are formed and consequently observed property changes can be correlated with the fundamental lattice defects. Where very accurate experimental techniques are available the electron energy can be lowered to the stage at which displacement events just begin to occur.

The presence of defect pairs is conveniently detected in both resistivity and internal friction measurements. In resistivity measurements it is usual to bombard thin foils uniformly with electrons. For specimens of length L, width W and thickness t, and for a homogeneous distribution of induced damage, the resistivity change per electron per cm^2 is given by

$$\frac{\Delta\rho}{\rho} = \frac{Wt}{Ln}\Delta R \tag{6.11}$$

where ΔR is the measured resistance change and n is the current density. Measured values of $\Delta\rho/n$ for several metals at 20 K are plotted as a function of average beam energy in the range 0·5 to 1·4 MeV in Fig. 6.13. The average threshold energies and resistivities of Frenkel pairs obtained by analysing such curves in terms of the theoretical model for displacement events are given in Table 6.2.

In internal friction studies, it is convenient to measure the logarithmic decrement of the free decay of oscillations at fixed frequency. For frequencies in the kilohertz range the internal friction in metals is predominantly due to dislocation motion. The vibrations which characterize this anharmonicity correspond to oscillations of dislocation segments of critical length, alternately bulging out on to atomic planes on either side of their original position. The critical dislocation length is determined solely by considerations of minimum line energy. Consequently, when crystals are irradiated at low temperature, dislocation pinning by interstitials decreases the amplitude rather than the frequency of vibration. On annealing above the irradiation temperature the interstitials migrate to dislocation lines and enhance the extent of the pinning. Thus the logarithmic decrement is further decreased. Usually several peaks occur in the curves obtained by plotting internal friction versus annealing temperature.

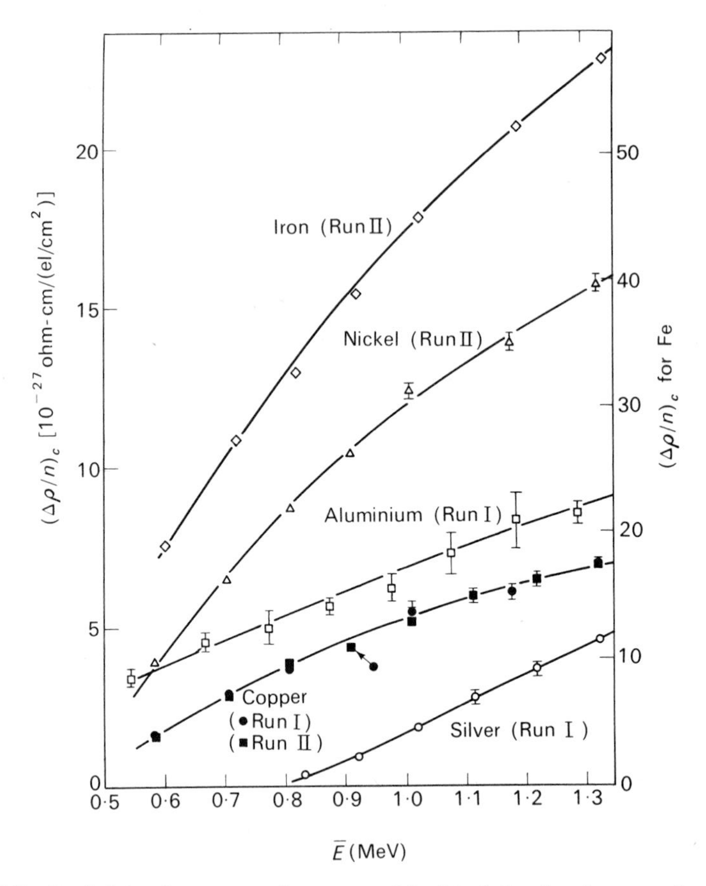

Fig. 6.13 Resistivity changes per electron cm^{-2} in thin foils of various metals uniformly bombarded with a beam of fast electrons of mean energy $\bar{E}$. (After LUCASSON and WALKER. 1962. *Phys. Rev.* **127,** 485.)

Table 6.2 Displacement energy and resistivity ($\Delta\rho$) per Frenkel pair in e-irradiated metals

	Al	Au	Ag	Cu	Fe	Mo	Ni	Ti	W
$\Delta\rho$ $\mu\Omega$. cm 1% defect	3·4	—	1·4	1·3	12·5	4·5	3·2	4·2	—
Threshold energy (eV)	32	>40	28	22	24	37	24	29	>35

(After LUCASSON and WALKER. 1962. *Phys. Rev.* **127,** 485)

It is informative to compare the stages observed in the annealing of the internal friction with those observed in the annealing of the resistivity. Figure 6.14 gives such a comparison for electron-irradiated copper and molybdenum. At low temperatures the resistivity changes show three stages A-C which are correlated with the recombination of close-interstitial-vacancy pairs. At slightly higher temperatures two stages D and E are attributed to arise from the free migration of interstitials to vacancies, stage D being considered as a correlated recovery caused by the interstitial moving back to its own vacancy.[57] Stages D and E are also observed by internal friction: Lomer and Taylor[58] attribute these stages to dislocation pinning by freely migrating interstitials. At the onset of pinning interstitial migration occurs in the neighbourhood of the dislocation strain field, and this is controlled by the stress-assisted interstitial migration.

A measurement of the displacement energy may also be obtained by monitoring the electron energy dependence of the logarithmic decrement. The method in general gives larger values for E_D than the resistivity method. This confirms that the defect structure is best pictured to include a spectrum of Frenkel defects ranging between those which consist of close interstitial-vacancy pairs to interstitials which are found at large distances from their vacancies. Extra energy is lost in the displacement chain of the more widely separated Frenkel pairs, which consequently have a larger apparent displacement energy. Resistivity measurements give a lower activation energy since they monitor all Frenkel defects including the close-pairs, whereas the internal friction is sensitive only to the widely separated interstitial-vacancy pairs and so results in the anomalously high displacement energies.

Heavy particle irradiation results in a very complex damage structure. Even for copper, which has been extensively investigated, the exact nature of the radiation damage is incompletely understood. The damaged region per energetic particle may be expected to contain about 10^5 atoms, there being interstitial defects around the periphery and vacancies in the centre as shown in Fig. 6.15. Such defects may be studied in a variety of ways and the lattice parameter, density and resistivity methods above have all been used. Electron microscopy has proved especially fruitful since the displacement cascade which results from particle damage certainly involves irradiation-induced diffusion of defects. The nature of defect clusters will depend strongly upon whether the displacement cascade mechanism predominates (vacancy clusters) or whether the diffusion mechanism predominates (intersitital clusters).

Apparently both vacancy clustering in the cascades and interstitial cluster formation by diffusion occur in metals during fast neutron irradiation. Either or both types of defects are observed by electron microscopy depending upon the fast neutron spectrum, dose rate and total dose as well as irradiation temperature. The clusters are revealed as black-white images, the contrast depending not just upon the nature of the cluster itself but upon its depth in the foil. There is an advantage in using heavy ion bombardment, for example,

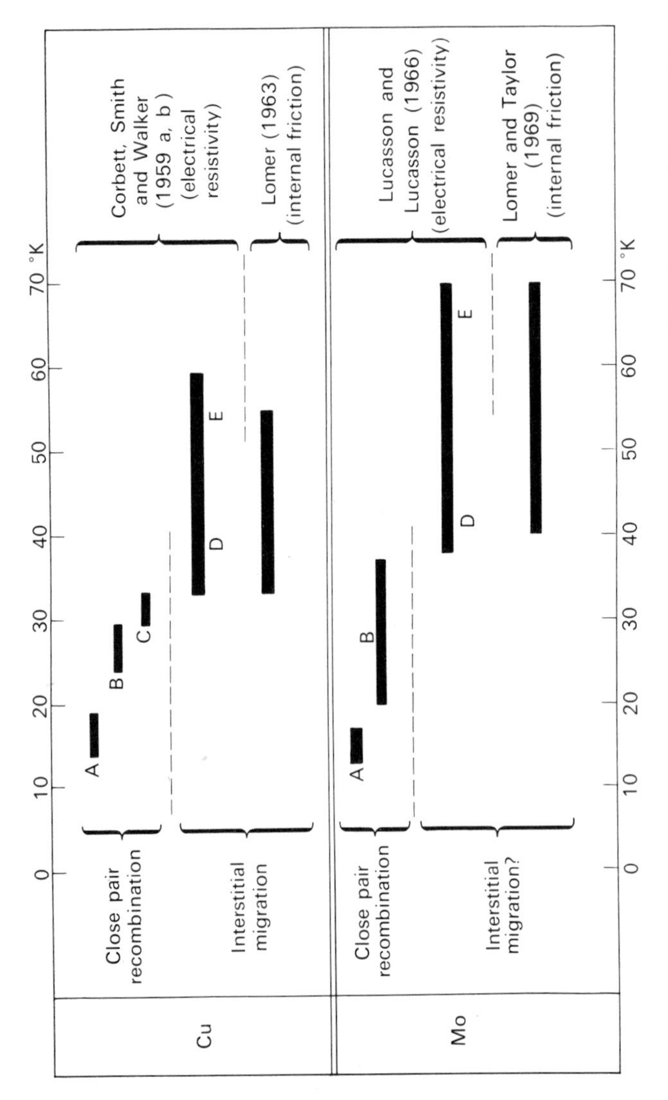

Fig. 6.14 Schematic annealing stages in electron irradiated Cu and Mo. (After LOMER and TAYLOR[58].)

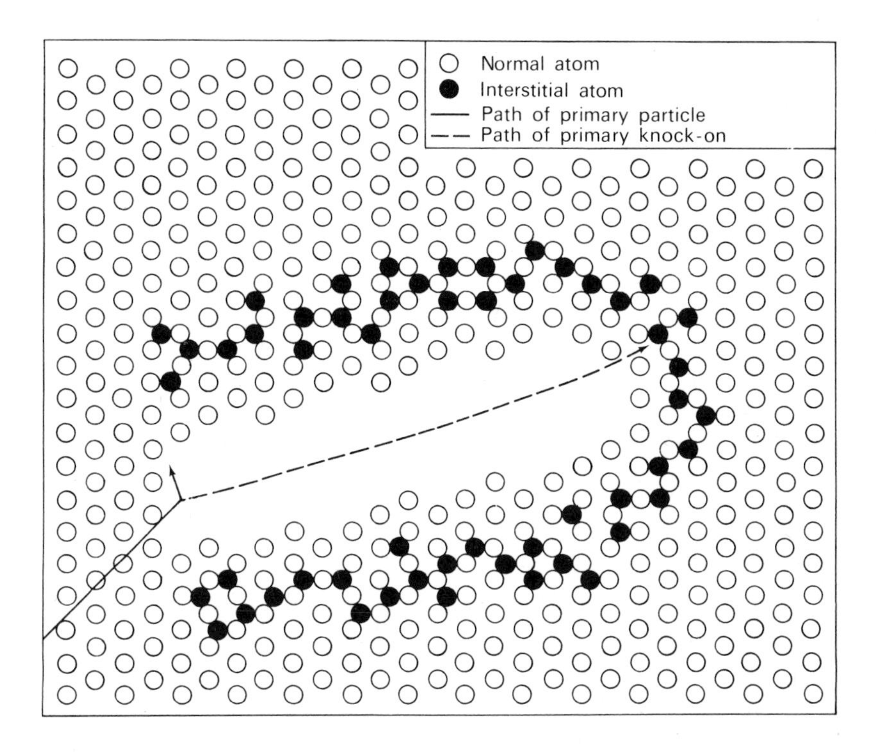

Fig. 6.15 Schematic representation of interstitials around a multiple vacancy during production of a displacement spike. (After BRINKMAN, J. A.), 1953, *J. App. Phys.* **25**, 961

80 keV Au^+ ions, since the damage is confined to a relatively thin region of the crystal near the surface. It is now well established that for many metals low doses (10^{15} ions m^{-2}) of particle energies in the range 15 to 150 keV, large displacement cascades are produced in which vacancy clusters are formed. The heavier the ion used the more exaggerated are the cascade effects. At very high dose (10^{19} ions m^{-2}) interstitial clusters predominate since the overlap of the outer regions of the cascade enhances the diffusion mechanism.

GENERAL REFERENCES

FRIEDEL, J. 1964. *Dislocations.* Pergamon Press, London.
SMALLMAN, R. E. 1961. *Modern Physical Metallurgy.* Butterworths, London.
THOMAS, G. and WASHBURN, J. (Eds.) 1963. *Electron Microscopy and the Strength of Crystals.* Wiley, New York.
THOMPSON, M. W. 1968. *Defects and Radiation Damage in Metals.* Cambridge University Press.
VAN BUEREN, H. G. 1960. *Imperfections in Crystals.* Interscience, New York.

7

Interactions betweeen Dislocations and Other Defects

In this final chapter some interactions between various crystal defects are discussed, especially insofar as they may influence the mechanical strength of solids. Before discussing in detail the mechanisms of such strengthening, some comments upon both the intrinsic strength and sources of slip in real crystals are appropriate.

7.1 The Peierls-Nabarro force

Crystalline solids are expected to have maximum shear strengths in the range 0·03 to 0·1 μ, μ being the elastic modulus. Such high strength, although characteristic of atomic bonding, is rarely attained in practice because real solids are never free from imperfections. In ductile materials, the yield stress actually measures the stress required to cause existing dislocations to glide. In Fig. 1.6 it can be seen that the equilibrium dislocation structure is symmetrical. It follows that the stress required to initiate slip depends upon the unsymmetrical configuration achieved during glide. This stress is usually referred to as the *Peierls-Nabarro* or *lattice-friction stress*. According to Cottrell[2] consideration of the unbalanced forces acting on a dislocation during slip gives the yield strength σ_y as

$$\boxed{\sigma_y = \mu \exp(-2\pi w/b)} \qquad (7.1)$$

where the dislocation width w is related to the theoretical tensile strength σ_t by

$$\boxed{w = \mu/2\pi(1 - \eta)\sigma_t} \qquad (7.2)$$

where η is Poisson's ratio. Thus materials with wide dislocations have relatively low yield strengths and are regarded as intrinsically soft. The *free electron bond* which typifies most close-packed metals favours intrinsic softness because it is non-directional and can tolerate relatively wide dislocations on the closepacked planes. In ionic solids the dislocations are narrow because the operative slip system tends to be the one which avoids bringing ions of like charge together. The strongly-directional covalent bond in semiconductors also tends to produce narrow dislocations. Consequently both ionic crystals and semiconductors are intrinsically hard at low temperature.

Intrinsically hard solids are expected to soften rapidly with increasing temperature. Since the dislocations are narrow, atoms at the dislocation core need only jump a distance equal to the Burgers vector to move the dislocation forward by one spacing. This is analogous to atomic diffusion and is thermally activated. Thus we expect the yield stress to decrease according to

$$\boxed{\sigma_y = A \exp(-E/kT)} \tag{7.3}$$

where A is a constant and E is the activation energy required for the thermal motion of the dislocation. Such an exponential decrease in mechanical strength has been observed for a number of hard materials notably MgO, LiF, TiC and VC.[59] Intrinsically soft materials are little affected by temperature, since lattice resistance to the motion of partial dislocations is very weak.

The measured yield stresses of various cryrstalline solids are plotted as a function of the ratio T/T_m in Fig. 7.1, T_m being the melting temperature. Although the mechanical strengths of the close-packed metals nickel and copper hardly change as a function of temperature, both silicon and aluminium oxide show very marked softening at $T > 0{\cdot}5T_m$. Hexagonal close-packed metals are expected to be particularly interesting since the Peierls-Nabarro force for dissociated dislocations moving on the basal planes is much less than for narrow dislocations moving on non-basal planes. Figure 7.1 shows that this expectation is realised for magnesium, the critical resolved shear stress and its temperature dependence being greater for glide on non-basal planes than for glide on the basal planes. Body-centred cubic metals, including the ordered intermetallic phases NiAl, CuZn and AgMg, also show hardening characteristics as a function of decreasing temperature, although the extent to which the hardening is intrinsic is uncertain. The transition metal carbides are typical hard materials and decrease in strength at elevated temperature (see §7.5). However, their strength increases markedly with increasing carbon content, being greatest for stoichiometric compounds. This is contrary to observations on other compounds such as rutile (TiO_2), where deviations from the stoichiometry increase the flow stress. This is because the mode of deformation of the transition metal carbides is controlled by movements on the close-packed transition metal sublattice, the carbon atoms contributing only to the dislocation mobility through the lattice friction stress.[59] Thus the greater the concentration of carbon vacancies the greater will be the decrease in flow stress, at least in the absence of vacancy ordering.

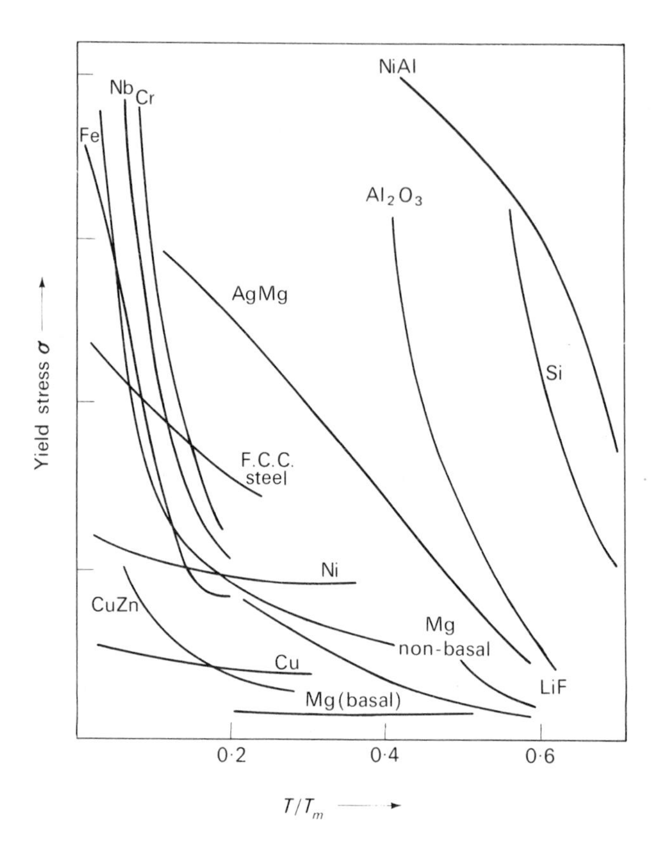

Fig. 7.1 Showing the temperature dependence of the yield stress of some crystalline solids. (After COTTRELL. 1964. *The Mechanical Properties of Matter*. Wiley, New York.)

7.2 Sources of slip in crystals

The processes by which dislocations get into crystals in large concentrations during deformation are important in developing a theory of crystal plasticity. Dislocations formed during crystal growth are usually locked by interactions with other defects. Consequently, such dislocations must be released by the applied stress or new dislocations must be created elsewhere in the crystal. It is possible for dislocations to be nucleated in regions of defect-free crystal. Since such *homogeneous nucleation* requires stress levels of order $\mu/30$,[2] this constitutes an unlikely source of new dislocations. Dislocations may also be created by *heterogeneous nucleation* at inclusions or precipitates. Particles only 1000 Å in diameter can cause dislocation nucleation since they act as stress concentrators as a result of differential thermal contraction between particle and matrix. Figure 7.2 is an example of prismatic dislocation loops being punched out from chromium oxide precipitates in chromium subjected to 150 k bars of hydrostatic pressure. However most probably the vast bulk of dislocations required for extensive plastic deformation result from *regenerative multiplication* through *multiple-cross-glide*, or the operation of *Frank-Read sources*.

According to the Frank-Read mechanism (see Fig. 7.3) a short length of dislocation locked by obstacles A and B will bow out on its slip plane under the action of an applied shear stress. As the stress is increased to a maximum the radius of the dislocation line becomes semi-circular. Beyond this the dislocation is unstable and continually expands until the two parts of the line, spiralling about A and B respectively, approach each other in opposite directions. When these dislocation segments meet at C they annihilate one another, since their Burgers vectors are of opposite sense, so producing a dislocation loop and regenerating the original dislocation AB. Whilst the dislocation loop is now free to expand on the slip plane the source dislocation is available to repeat the process again. This mechanism can obviously be activated indefinitely so producing intense slip on its slip plane. Notice that if there is an obstacle to slip on the slip plane, a grain boundary for example, the dislocations nucleated by the Frank-Read source will tend to pile up against the obstacle. An equilibrium situation will eventually be brought about between the applied stress and the back stress exerted by the

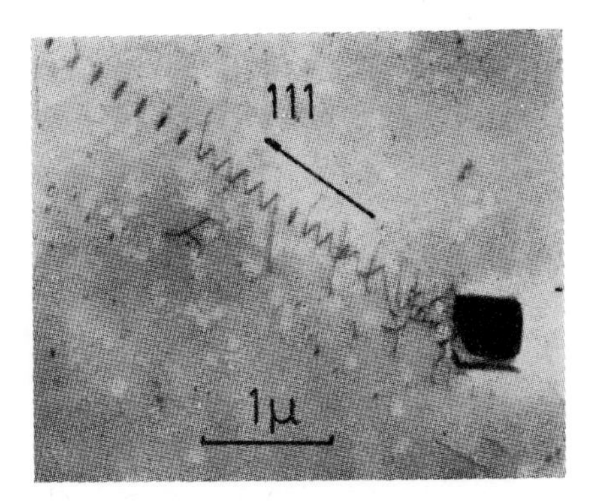

Fig. 7.2 Prismatic dislocation loops in chromium. (Reproduced by kind permission of Dr. A. BALL from BALL and BULLEN 1970, *Phil. Mag.*, **22**, p. 304.)

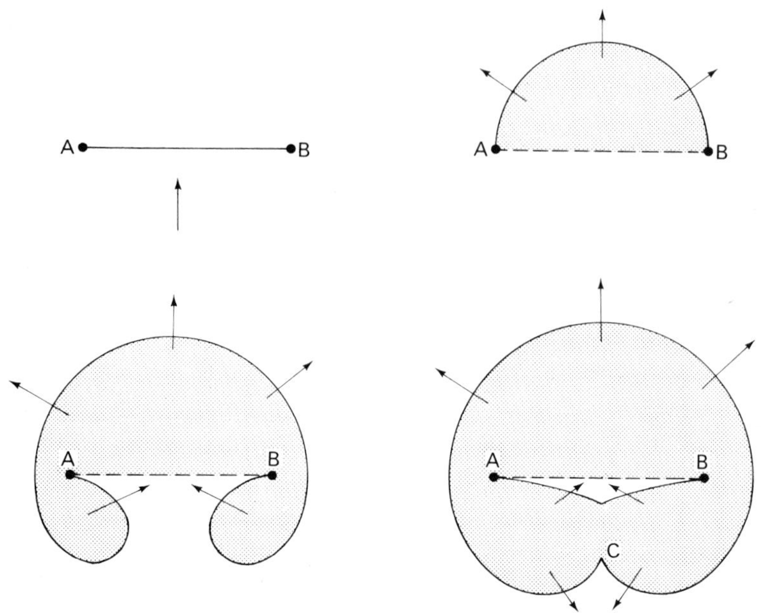

Fig. 7.3 The operation of a Frank-Read source.

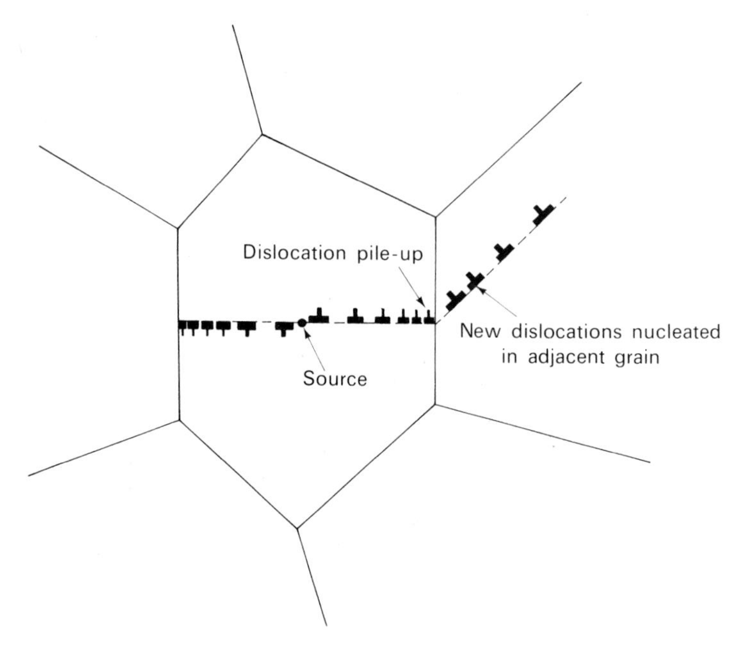

Fig. 7.4 Dislocations nucleated at a Frank-Read source pile-up at grain boundaries. When the stress on the slip planes of neighbouring grains is high enough dislocation sources in the new grains are activated.

piled-up dislocations, and the source will cease to operate. A schematic representation in Fig. 7.4 shows that the dislocations become increasingly closely spaced towards the grain boundary.

It is apparent from its mode of operation that the Frank-Read source will produce a set of concentric and coplanar dislocation loops. Thus the model predicts very sharp glide bands. Since plastic deformation results in broad diffuse glide bands additional dislocation sources must operate. The process of multiple cross-glide is believed to predominate in crystals with narrow dislocation lines. Such a process is indicated schematically in Fig. 7.5. The first step in this mechanism is the formation of the jogs *EC* and

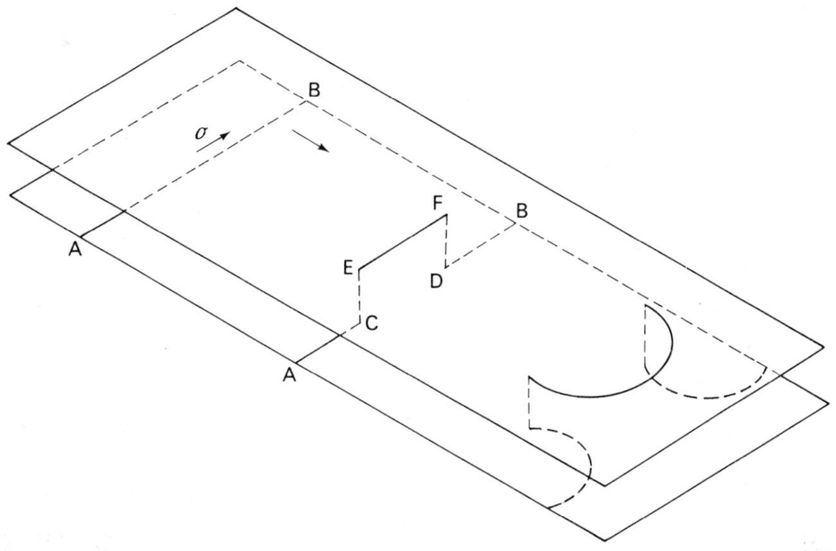

Fig. 7.5 A cross-glide-multiplication mechanism. (After GILMAN and JOHNSTON. 1962. *Solid State Physics*. **13**, 147, Academic Press, New York.)

FD in the screw dislocation *AB*. Since these jogs constitute edge dislocations, they cannot move in the direction of the applied stress and are consequently immobile. Thus the dislocation segments *AC*, *EF* and *DB* can bow out on their respective slip planes and hence *EF* and *CD* may behave as Frank-Read sources on two parallel planes. If cross-slip is a relatively frequent occurrence (as in ionic solids and metals with high stacking fault energy) the Frank-Read sources may again cross-slip before a loop is completed. Thus by repeated cross-slip from one slip-plane to another a thick glide band is formed: there will be only one continuous dislocation line with sections lying on many parallel slip planes and with jogs connecting the various sections. Consequently when multiple-cross-glide is important there are no concentric loops: instead the many dislocation segments will result in the broad, diffuse glide bands observed experimentally.[60]

7.3 Interactions of point defects with dislocations

In general the most important interactions between point defects and dislocations occur as a result of their respective elastic strain fields. In the case of impurity atoms in a random solid solution only moderate hardening is observed, since on average there are equal numbers of solute atoms on either side of the dislocations resulting in an equal interaction from both sides. Thus this interaction can be regarded as a minor perturbation on the intrinsic friction stress of the pure crystal. However, of greater significance, especially for materials of technological importance, are interactions which result in impurity segregation on dislocation lines, impurity clustering and precipitation. We discuss first the origin and magnitude of the elastic interaction between point defects and dislocations: other interactions are discussed in less detail.

The elastic interaction

Figure 1.8 shows that whilst the strain field around a screw dislocation is symmetrical, having shear components only, the strain field around an edge dislocation is unsymmetrical and possesses both hydrostatic and shear components. When a solute atom is substituted for a solvent atom the site which it occupies will almost certainly be the wrong size and possibly also the wrong shape for it. Consequently the lattice will relax around the solute atom; according to linear elasticity theory a spherically symmetric atom will distort the lattice equally in all directions. (N.B. In principle the same considerations also apply for the strain fields around vacant lattice sites.) Thus there is a "size effect" interaction between the stress fields of the inclusion and the dislocation, the magnitude of which was first calculated by Cottrell and Bilby.[61] The solute atom of radius $r_0(1 + \epsilon)$, ϵ being the 'misfit' parameter, represented by an elastic sphere of identical radius, is "squeezed" into a spherical hole of radius r_0 in an infinite elastic medium representing the crystal. Both spherical inclusion and matrix are assumed to have the same elastic constants. The isotropic strain field of the solute atom interacts only with the hydrostatic components of a strain field. Thus to first order there is no size effect interaction between a point defect and a screw dislocation. Consider now the elastic interaction between the stress field of a long, straight edge dislocation along the z-axis and a spherical inclusion situated at a point P with polar co-ordinates (R, θ). The interaction energy may be written as

$$E_1 = p\,\Delta v$$

where p is the hydrostatic pressure of the edge dislocation's stress field and Δv is the volume change caused by the atom. The dilatational strain of the edge dislocation, Θ, at (R, θ) according to Cottrell[2] is,

$$\Theta = \frac{-b}{2\pi}\left(\frac{1-2\eta}{1-\eta}\right)\frac{\sin\theta}{R} \tag{7.3}$$

where η is Poisson's ratio. Thus for a material with bulk modulus $K = 2(1+\eta)\mu/3(1-2\eta)$ and since $p = K\Theta$ we obtain,

$$p = \frac{-\mu b}{3\pi}\frac{(1+\eta)}{(1-\eta)}\frac{\sin\theta}{R}$$

Substituting $\Delta V = -4\pi r_0^3\epsilon$, the interaction energy is

$$E_1 = \frac{4\mu b}{3}\left(\frac{1+\eta}{1-\eta}\right)\frac{\epsilon r_0^3 \sin\theta}{R} \tag{7.4}$$

This result applies both for spherical atoms ($r_0 \leqslant R$) and for large spherical precipitates. Notice that for a large atom, E_1 is positive above the slip plane when $0 < \theta < \pi$, and negative below the slip plane. Consequently a large atom is attracted to the expanded region of the dislocation core below the slip plane. Similarly a small atom (or vacancy) is attracted to the compressed region above the slip plane.

The case of carbon and nitrogen in body-centred cubic iron has occupied a central position in the development of the theories of yield point phenomena. For carbon in iron maximum binding should occur at $\theta = 3\pi/2$ and $R = r_0$. Using $r_0 = 2$ Å and $b = (a/2)\langle 111\rangle = 2{\cdot}5$ Å we obtain $E_1 = 1$ eV when the known values of ν, μ and ϵ are substituted in Equ. 7.4. For substitutional solid solutions of zinc in copper, $\mu = 4 \times 10^{11}$ dyn cm^{-2}, $\eta = 0{\cdot}36$, $b = 2{\cdot}55$ Å, $r_0 = b/2$ and $\epsilon = 0{\cdot}06$, and a much lower binding energy of only 0·13 eV is obtained. Consequently much weaker yield point phenomena are observed in the α-brasses than in mild steel. Similarly obtained values for a wide range of substitutional and interstitial impurities in silicon and germanium have been tabulated by Bullough and Newman.[62] In this case the misfit parameter $\epsilon = (r_0 - r_i)/r_0$, is defined in terms of r_0 and r_i the covalent radii of the host material and the impurity. The interaction energies for boron and phosphorus in silicon at 1200°C are respectively 0·79 eV and 0·18 eV.

So far the discussion of the size effect interaction with edge and screw dislocations has been somewhat idealized since it presupposes that,

(a) the crystal is elastically isotropic and that linear elasticity theory obtains,
(b) dislocations are long and straight and are in either the pure edge or pure screw configuration,
(c) a spherical inclusion is a good representation of the solute atom,
(d) the elastic constants of impurity atom and the matrix are identical.

The breakdown in linear elasticity theory near the dislocation core may be taken account of by asserting that

$$E_1 = \frac{4\mu b}{3}\left(\frac{1+\eta}{1-\eta}\right)\frac{\epsilon r_0^3 \sin\theta}{R+w} \tag{7.4a}$$

where w is the effective width of the dislocation. Using a more realistic value of $E_1 = 0{\cdot}45$ eV for carbon in iron (determined using internal friction methods) and by substituting for η, μ, b and r_0 in Equ. 7.4a we obtain $w = 4$ Å: this is a reasonable value for the width of undissociated dislocations in body-centred cubic metals. In face-centred cubic metals the dislocation width depends largely upon the stacking fault energy. According to §1.3 screw dislocations on close-packed planes dissociate into partial dislocations which, being partially edge and partially screw, are susceptible to interaction with spherically symmetric strains. Consequently, in materials with low stacking fault energy, screw dislocations still effectively interact elastically with spherical inclusions. In addition to this indirect locking of the screw component of the dislocation line, there is also a weak direct interaction. This arises because the out-of-register atomic planes near the centre of the dislocation force each other apart. Such second order volume expansions produce hydrostatic components in the stress fields of edge and screw dislocations which interact elastically with the inclusion. Fleischer[63] estimates that the interaction energy associated with this effect for a screw dislocation is

$$E_2 = \frac{2\mu r_0^3 \epsilon}{3\pi}\left(\frac{1+\eta}{1-\eta}\right)\left(\frac{b}{R}\right)^2 \tag{7.5}$$

For edge dislocations, where the first order size effect interaction is large, such a contribution may be neglected.

The shape of the solute atom may be important since deviations from spherical symmetry cause interactions with both shear and hydrostatic components of the dislocation stress field. Thus non-spherical atoms interact elastically with both edge and screw dislocations. Similarly interstitial atoms may introduce anisotropic lattice distortions. Although relatively unimportant in face-centred cubic crystals, this effect becomes important in hexagonal close-packed and body-centred cubic crystals. For example, carbon atoms dissolve interstitially in body-centred cubic iron, occupying sites of tetrahedral symmetry (Fig. 1.1). Thus the elastic interaction between carbon impurities and the elastic strain field of a dislocation is as strong for screw dislocations as for edge dislocations.

There is a further contribution to the elastic interaction between an impurity atom and a dislocation line which results from the different elastic moduli of the inclusion and the matrix. Fleischer[63] calculates the change in energy which results from this type of interaction, for both edge and screw dislocations. He finds that the relative contributions to the interaction energy from the size effect and modulus difference terms are approximately equal for the edge dislocation: for screw dislocations the contribution due to the modulus difference effect is dominant. The hardening effect due to vacant lattice sites has been calculated,[62] by treating the vacancy as an infinitely soft spherical inclusion. The interaction energy in this case takes the form

$$E_3 = \frac{-5\mu b^2 r_0^3}{\pi R^2} F(\eta, \theta) \qquad (7.6)$$

where r_0 is the radius of the vacancy and the function $F(\eta, \theta)$ depends upon whether the dislocation is edge or screw in character. Note that the *inhomogeneity* and size effect interactions are not necessarily in the same sense. Thus a large rigid atom is attracted to the dislocation by the size effect interaction and repelled from it by interaction due to the differences in shear modulus of impurity and matrix. However, for vacancies Equ. 7.6 always predicts attractive interactions between the vacancy and dislocation.

Other interactions

The ability of a solute atom to harden a metallic solid solution depends not only upon the degree of misfit relative to the matrix but also upon its relative valency. Consequently solute atoms with different valencies but identical radii harden a solid solution to different extents. This effect is due to an electrostatic interaction between the excess charge on the solvent atoms and the unsymmetrical charge distribution around edge dislocations. The charge associated with the dislocation arises from a change of Fermi level near the edge dislocation[64] given by

$$E_F = \frac{-4}{15} E_{\max} \,.\, \Theta$$

where Θ is the dilatation strain defined by Equ. 7.3. Consequent upon this local change in the Fermi level, some electrons redistribute themselves in order to equalize the Fermi level everywhere and there is a net flow of electrons from the compressed region of the lattice above the slip plane to the expanded region below the slip plane. Since the effective charge on an impurity atom of valency Z is given by $0{\cdot}075\ (Z - 1)$ electronic units[65] the interaction energy due to this electrical interaction is

$$E_4 = -0{\cdot}02(Z - 1)E_{\max}\,\Theta \qquad (7.7)$$

For zinc in copper $Z = 2$ and $E_{\max} = 7$ eV, hence $E_4 \approx 0{\cdot}02$ eV at the position of maximum binding. This is to be compared with the binding energy due to the elastic interaction of $E_1 = 0{\cdot}14$ eV. Similar calculations for other impurities from Groups II to V of the Periodic Table show that in copper the elastic interaction is stronger than the electric interaction by a factor which decreases from 7 for zinc to only 3 for arsenic.

Although the electrical interaction is of little significance in metals it may be important in semiconductors and in ionic crystals. Edge dislocations in semiconductors induce a weak electrical interaction with impurities which is essentially similar to that in metals. However, there is a second possible electrical interaction due to the "dangling bonds" along the dislocation core (see Fig. 5.15). In n-type crystals these bonds capture electrons from the conduction band until the dislocation line becomes surrounded by a cylindrical region of space charge. Even here the interaction is relatively weak, amounting to only $\sim$0·04 eV for a unipositively charged ion. The interaction is unlikely to be a major source of hardening above 300 K. This effect is not present in p-type material. Stronger electrical interactions are present in ionic crystals due to the absence of screening effects. Since jogs on dislocation lines carry an effective charge the interaction may also be very long ranged; in practice the electrical interaction may be stronger than the elastic interaction.

A last interaction to be considered here is the Suzuki or chemical interaction,[66] which applies only when dislocations can dissociate into partial dislocations bound together elastically by a stacking fault. Since in general the structure of the fault differs from that of the matrix, a concentration difference may be established between the two regions on account of their different thermodynamic potential. This can result in the fault energy being lowered when the impurity concentration in the faulted region exceeds that in the matrix. Again this is a very weak interaction, although being essentially temperature independent, it may be dominant at high temperature.

It is now evident that point defects in crystals may be strongly attracted to dislocations and that this attraction effectively reduces the dislocation mobility. The total binding energy of a point defect to a dislocation may be directly summed from the contributions discussed above, although such contributions are not necessarily in the same sense. Thus an impurity at a point P with polar co-ordinates (R, θ) experiences a total interaction energy $E(R, \theta)$ with a dislocation line at the origin, this interaction energy being a maximum at the dislocation core. At high temperature, where the impurities are mobile and equilibrium is established before impurities have occupied the sites at the dislocation core, the probability of an impurity atom entering and leaving the dislocation must be equal. This equilibrium state is then a Maxwellian atmosphere with a concentration distribution

$$C(R, \theta) = C_0 \exp\left(E(R, \theta)/kT\right) \tag{7.8}$$

where C_0 is the impurity concentration a long way from the dislocation. This is only true if the dislocation atmosphere remains dilute, otherwise the detailed occupation statistics of available sites must be taken into account. Thus an impurity atmosphere gathers around the dislocation line, the concentration of which decreases as we move out from the dislocation line. When $E_{max} \gg kT$ the atmosphere has passed from a dilute to a condensed state in which $C_{max} = 1$, corresponding to the atmosphere condensing into a single line of impurity atoms lying along the dislocation line at the position of maximum binding. The temperature, T_c, at which condensation occurs is obtained by substituting $C_{max} = 1$ in Equ. 7.8; thus

$$T_c = E_{max}/k \log_e (1/C_0) \tag{7.9}$$

Thus for a dilute alloy containing 1% impurity and in which $E_{max} \approx \frac{1}{8}$ eV, the impurity atmosphere condenses at $T_c \approx 300$ K. Consequently at temperatures below T_c dislocation lines are very strongly bound or "locked" by the impurity atmosphere.

7.4 Defect interactions and the deformation of solids

The shape of the stress-strain curve

Some typical stress-strain curves are shown in Fig. 7.6. In all cases the initial deformation is elastic, stress (σ) and strain (ϵ) being linearly related. Other features shown in Fig. 7.6 are dependent upon crystal orientation, purity, temperature and strain rate. This last experimental variable depends largely upon the type of tensile machine used. For many experiments *hard* tensile machines are preferred since they undergo only small elastic distortions at high stress, as a consequence of which they are particularly responsive to yield points.

For single crystal specimens of pure face-centred cubic metals there are three well-defined regions in the stress-strain curve (Fig. 7.6a). Stage I, or easy-glide, which corresponds to deformation on a single glide plane, can account for shear strains up to 40%. However, the extent of Stage I hardening differs from metal to metal, being greatest for metals with low stacking fault energy (e.g. Cu, Ag, Au) since cross-slip is difficult to induce on account of the widely separated partial dislocations. When the stacking fault energy is high (Al, Ni), dislocations are essentially undissociated and cross-slip occurs readily: under such circumstances Stage I is limited to about 5% plastic strain. Impurities may also influence the extent of Stage I hardening. When present in solid solution impurities increase the extent of Stage I consequent upon a decrease in the stacking fault energy. Impurities which are present as a finely dispersed second phase decrease the extent of Stage I, since precipitation encourages local glide on other slip planes. In the Stage II or linear hardening region, the slope of the stress strain curve is up to 10 times greater than in Stage I. However, the inception and extent of Stage II hardening are strongly temperature dependent. At room temperature aluminium crystals show almost no Stage II hardening, whereas at 77 K a well defined Stage II starts after about 5% strain in Stage I. During Stage II there is a progressive development of slip on secondary systems so that

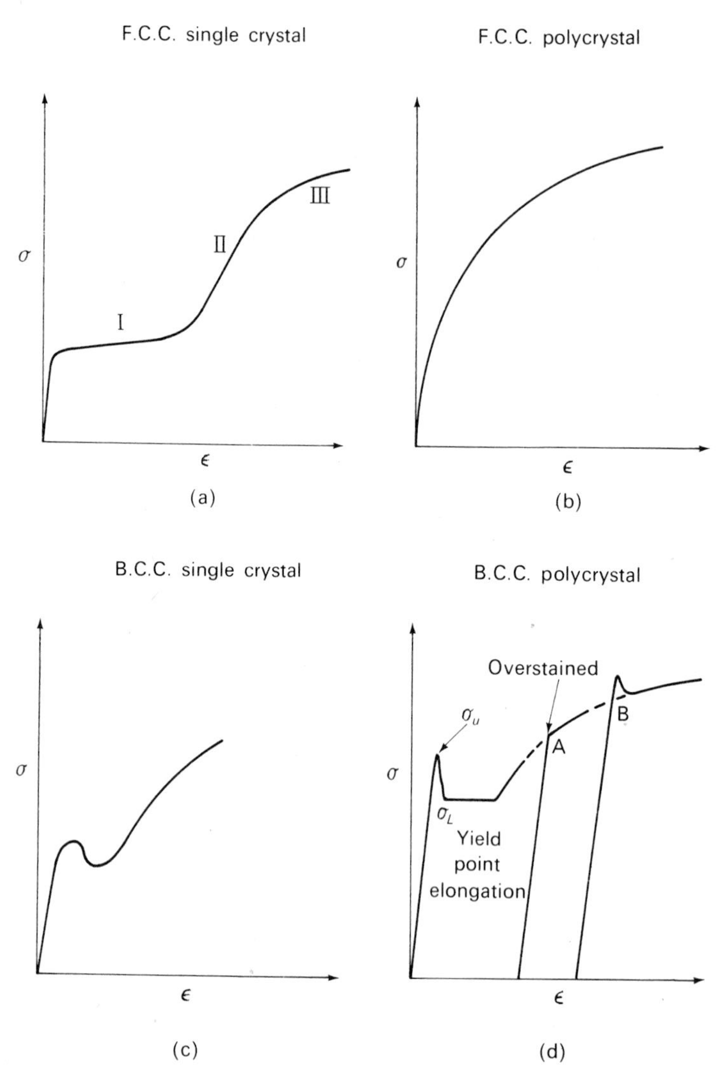

Fig. 7.6 Some typical stress–strain curves in metals.

both sessile dislocations (Lomer-Cottrell barriers, §7.6) and dislocation networks are being formed. During Stage III hardening, the rate of hardening decreases mainly as a result of the stress level being high enough to allow cross-slip of screw dislocations across the Lomer-Cottrell barriers. This process is thermally activated and that the onset of Stage III or parabolic hardening is strongly influenced by temperature will occasion no surprise.

Consequently, in aluminium at room temperature, Stages I and III are almost merged. Polycrystalline specimens do not show an easy glide stage: the deformation is characterized by a parabolic stress-strain curve equivalent to Stage III in single crystals (Fig. 7.6b).

Single crystal and polycrystalline samples of body-centred cubic metals exhibit essentially similar stress-strain curves (Figs. 7.6c and d). The stress strain curve increases linearly with strain up to the *upper yield point* (σ_u), at which the applied stress exceeds that necessary to maintain the strain rate in the crystal imposed by the machine. There is a corresponding *yield drop* to the *lower yield point* σ_L, which is succeeded by the yield point elongation, during which strain increases at almost zero stress. With continued increasing stress the specimen *work hardens* at a fairly uniform rate. Two additional features of stress-strain curves are to be noted. If a specimen is deformed to the *over-strained* or work hardened state at A in Fig. 7.6d, unloaded and immediately reloaded no yield drop would be observed. Furthermore plastic deformation does not restart until the stress at which the specimen was originally unloaded is reached. Thus the dislocation motion involved in plastic deformation is essentially irreversible. If, however, an overstrained specimen is allowed to "age" for some period before being reloaded the yield point returns as at B in Fig. 7.6d. This process of *strain-ageing* is frequently speeded up by annealing at some temperature higher than the measuring temperature.

Theories of yield points

The early theories of yield points[2,61] attributed the well-defined yield-points in metals to pre-existing dislocations becoming suddenly released from impurity atmospheres by the applied stress. The upper yield point σ_u is thus the stress required to overcome the attractive interaction between impurity and dislocation discussed in §7.3. Similarly the lower yield point σ_L is the stress required to maintain the machine-imposed strain rate in the presence of a high concentration of released dislocations. Strain-ageing is accounted for by asserting that the impurity atoms have diffused back to the freed dislocation lines during the ageing period.

Despite its undoubted success when applied to the hardening of Fe by C, the atmosphere-locking theory is not generally acceptable for several reasons. This first became apparent when Gilman and Johnston[60] demonstrated that in lithium fluoride many *new* dislocations are generated during yielding. These authors accounted for the new dislocations in terms of the multiple-cross-slip mechanism discussed in §7.2. Furthermore it is evident from experiments on alloys that although yield points may develop as the alloy addition increases, there is also a general increase in the level of the stress-strain curve as a whole. This indicates that the main contribution to the stress level required for plastic flow to occur is the increased friction stress on moving dislocations, rather than the binding interactions discussed in §7.3. The modern viewpoint predicts that yield points of varying intensity are a normal characteristic of metals and alloys, the absence of such phenomena being due to bad experimental technique. However, Cottrell[67] has pointed out that when the rate of work hardening over the critical strain range is high the plastic instability at the yield point is difficult to detect. Consequently it is easier to detect yield points in face-centred cubic single crystals (since the rate of hardening in Stage I is low) than in polycrystalline samples. Thus it becomes necessary to devise a theory which will lead to yield point phenomena irrespective of whether there are impurities present or not.

A simple theory which takes account of these features was shown by Gilman and Johnston[60] to require only two essential criteria for the occurrence of sharp yield points: 1) The initial dislocation density must be initially small and increase rapidly during the yield drop. (2) Although the stress and dislocation velocity are directly related, the dislocation velocity must not initially increase too rapidly with increasing stress. According to Gilman and Johnston's theory[60] there is an initial concentration N of unlocked dislocations present in the crystal. Thus during the elastic region the strain is given by

$$\epsilon = Nb\bar{S}$$

where b is the dislocation Burgers vector and $\bar{S}$ is the mean displacement of the dislocation. Consequently the strain rate of the specimen is given by

$$\dot{\epsilon} = Nb\bar{V} \tag{7.10}$$

$\bar{V}$ being the mean dislocation velocity. Evidently if there is a rapid multiplication of dislocations at the upper yield point then the same strain rate can be maintained at a lower dislocation velocity. Since the dislocation velocity is directly related to the applied stress, the stress level decreases to σ_L. If the applied stress σ and dislocation velocity are related by $\bar{V} = k\sigma^m$ then from equation 7.10

$$\frac{\sigma_u}{\sigma_L} = \left(\frac{N_L}{N_u}\right)^{1/m} \tag{7.11}$$

since the strain rate at the upper and lower yield points are identical. Thus the extent of the yield drop is determined by the parameters N_u and m. A family of deformation curves calculated from Equ. 7.11 for constant N_u and varying m are shown in Fig. 7.7. It follows that for small values of m the ratio σ_u/σ_L is large and a pronounced yield drop is expected. When $m \rightarrow 100$ the yield drop is likely to be too small to be detected. Thus a general unpinning of dislocations is unnecessary in order to predict a yield drop.

Yield points have now been observed in a wide range of crystalline solids. Lithium fluoride, germanium and silicon all have relatively small values of m and consequently the observed yield points are well developed provided that the density of mobile dislocations present at the upper yield point is less than $\sim 10^4$ cm^{-2}. However, the most important example of the yield point occurs in soft iron or mild steel, where the very strong plastic instability

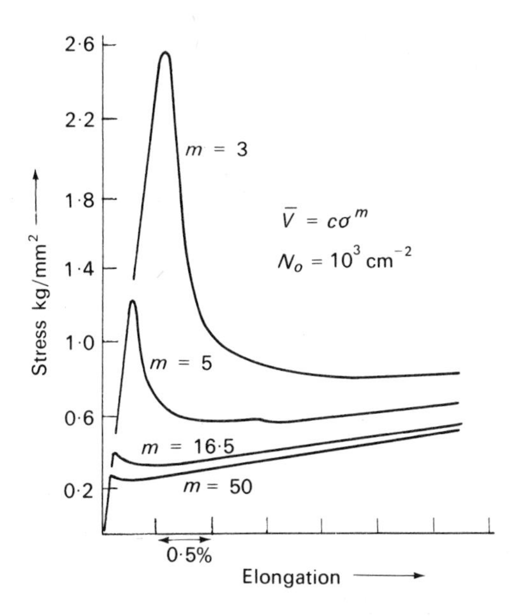

Fig. 7.7 Showing calculated stress–strain curves assuming variations in the interdependence of dislocation velocity on stress. (After JOHNSTON. 1962. *J. App. Phys.* **33,** 2716.)

at the yield point has traditionally been assigned to the dislocation locking by small concentrations of interstitial carbon or nitrogen. Now in body-centred cubic iron, $m = 35$, and thus the initial density of mobile dislocations must be small. Actually the yield point is substantial only when the value of N_u is less than about 10^3 cm^{-2}. Although the dislocation density in annealed iron greatly exceeds this, both edge and screw dislocations are strongly bound by interactions with interstitial impurities. In fact the impurities are present as a Cottrell atmosphere below the critical temperature given by Equ. 7.9. Since $E_{max} = 0{\cdot}5$ eV for interstitial carbon in Fe and assuming $C_0 \approx 10^{-4}$ we find that $T_c = 700$ K. Consequently we expect to find evidence of yield drops in mild steel at all temperatures below about 700 K. Above 700 K the impurities no longer constitute an impurity atmosphere so that N_u is greatly enhanced, there being a concomitant decrease in the yield drop. In agreement with this suggestion experimental results show that no upper yield point is observed in mild steel above 700 K. Other body-centred cubic metals show similarly pronounced plastic instability at the upper yield point.

Although yield point phenomena occur to a lesser extent in face-centred cubic metals, they are nevertheless well developed. This is shown in Fig. 7.8 for a series of alloys of zinc dissolved substitutionally in copper. The interaction energy between zinc atoms and dislocations in copper is relatively

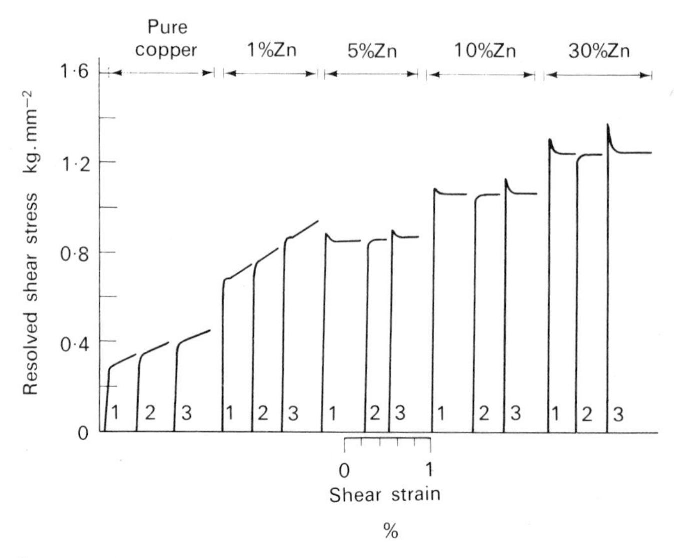

Fig. 7.8 Stress–strain curves on copper and brass crystals grown in argon, all strained at room temperature. Curve 1—first loading, Curve 2—immediately afterwards, Curve 3—after 2 hours at 200°C. (After ARDLEY and COTTRELL. 1953. *Proc. Roy. Soc.* **A219,** 328.)

weak ($\approx \frac{1}{8}$ eV). Consequently as much as one atomic percent of zinc is required to effect atmosphere-like condensation of zinc atoms onto dislocation lines in copper at 300°C. Accordingly, Fig. 7.8 shows that an alloy of Cu − 1% Zn has a sharp yield point. The analogy with mild steel is completed by observations on crystals in the overstrained and strain-aged condition. Although the magnitude of the yield drop obviously increases with increasing zinc content, the more noteworthy consequence of alloying is that the overall level of the stress-strain curve has been increased by a factor of 4. This does point to the change in friction stress experienced by the mobile dislocations in alloys being the dominant effect rather than the dislocation locking effect.

According to Fleischer[63] the frictional force is sensitive to both atomic size differences, and to differences in the elastic properties between solute and solvent. The interaction resulting from the modulus difference effects is given by

$$E_5 = \frac{\mu b^2 r_0^3 \epsilon_\mu}{6\pi R^2} \tag{7.12}$$

ϵ_μ defines the modulus difference between solvent and solute according to

$$\dot{\epsilon}_\mu = \frac{\mu^{-1}\, \mathrm{d}\mu/\mathrm{d}c}{1 - (2\mu)^{-1}\, \mathrm{d}\mu/\mathrm{d}c} \tag{7.13}$$

where μ, b, r_0 and R have their usual significance and c is the atomic concentration of solute. The rate of hardening as a function of solute concentration can then be discussed in terms of a combined mismatch parameter ϵ_c such that

$$\epsilon_c = \epsilon - 3\epsilon_\mu$$

where ϵ is the size difference parameter discussed earlier and ϵ_μ is the modulus difference parameter defined by Equ. 7.13. Fleischer plotted $d\sigma/dc$ versus the combined parameter ϵ_c for a wide range of copper-base alloys as we show

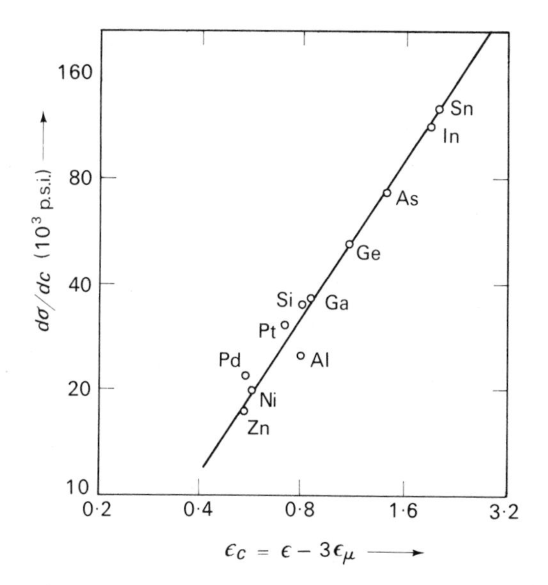

Fig. 7.9 Relation of the solid solution strengthening to combined mismatch parameter. (After FLEISCHER. 1963. *Acta Met.* **11**, 203.)

in Fig. 7.9. The results show that both effects substantially contribute to solid solution strengthening by alloying. Furthermore, the contribution of the two effects to the yield stress will differ according to the particular solvent. For the particular case of Cu–Zn alloys discussed above the size difference effect amounts to only about 10% of the yield stress increase. Where large size differences are apparent each interaction accounts for approximately half the increase in flow stress.

Yield points in polycrystals

Plastic deformation in body-centred cubic metals does not occur homogeneously at the upper yield point. Instead slip is nucleated in regions of high stress concentration, from which it spreads out through the rest of the material. Evidence for the heterogeneous nucleation of the plastic zone is provided by observations of *Lüders bands*, surface markings which occur on the specimen when the upper yield point is reached. These bands of plastic deformation spread along the specimen during the yield point elongation until the whole length of the specimen is deformed. The deformation in face-centred cubic and hexagonal close-packed metals is much less heterogeneous, at least during Stage I deformation.

In polycrystalline specimens of body-centred cubic metals the Lüders band is nucleated in a single grain. The high stress level at the upper yield stress causes an avalanche of mobile dislocations to be released on the slip plane. The grain boundaries act as barriers to slip: as dislocation pile-ups are produced the effective stress on the slip planes of neighbouring grains is increased. Eventually the stress is sufficiently high for mobile dislocations to be nucleated in neighbouring grains as indicated in Fig. 7.4. This process is then repeated again and again in other grains as the Lüders band propagates through the specimen. It is evident, therefore, that the grain size will be an important variable in the yield point variation of polycrystalline samples. It is relatively easy to show that the lower yield stress σ_L is given by the Petch[68] equation

$$\sigma_L = \sigma_i + k_y d^{-n} \tag{7.14}$$

where σ_i is the lattice friction stress, k_y is related to stress required to propagate slip across a boundary, and d is the grain diameter. The index n is usually $\frac{1}{2}$ in body-centred cubic metals, but is less well defined in face-centred cubic structures. However, it is generally observed that metals are stronger when fine-grained than coarse-grained. Figure 7.10 shows that Equ. 7.14 is obeyed for a low carbon steel, niobium and neutron irradiated copper at various temperatures.

The lattice friction stress term in Equ. 7.14 is of course independent of grain size, its magnitude being obtained from the intercepts in Fig. 7.10. Clearly the friction stress is strongly temperature dependent, increasing substantially as the temperature is lowered. The friction stress can arise from any type of obstacle on the slip plane which limits the dislocation mobility. Thus the value of σ_i in neutron irradiated copper is greater the longer the exposure to the neutron flux. Now according to the earlier discussion (§7.1) the temperature dependence of σ_i is expected to be weak in face-centred cubic metals. Thus the increased friction stress is attributed to dislocations moving across the slip plane having to overcome the resistance of the small clusters of vacancies and interstitials observed by electron microscopy.

Interpretation of the k_y term is less straightforward. It can be seen in Fig. 7.9b that for low carbon steel, k_y is somewhat temperature dependent. This would lend support to the earlier contention that k_y represents the stress required to unlock pinned dislocations in neighbouring grains as a consequence of a dislocation pile-up at grain boundaries. On the other hand

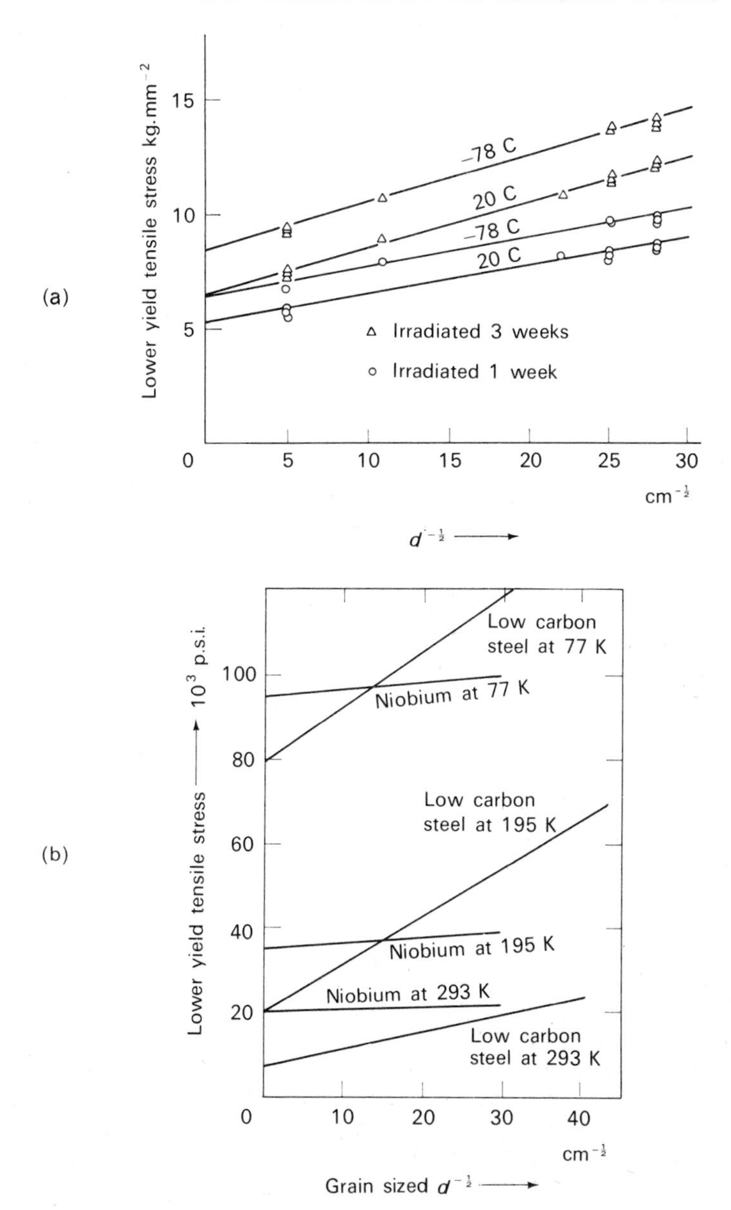

Fig. 7.10 Showing the variation in lower yield stress with grain size and temperature for (a) copper irradiated for different periods in a reactor (after ADAMS and HIGGINS. 1959. *Phil. Mag*. **4,** 777), and (b) low carbon steel and niobium. (After HULL and MOGFORD. 1958. *Phil. Mag*. **3,** 1213.)

k_y appears to be almost temperature independent in niobium and in n-irradiated copper, implying that dislocation locking is temperature independent. This dilemma may be resolved when the sizes of the dislocation-locking agents are properly taken into account. For example, if locking is in the classical Cottrell-Bilby sense then thermal fluctuations are certainly capable of helping to throw the dislocation loop forward so freeing it from its atmosphere. Thus k_y might well be temperature dependent. However this mechanism is unlikely since there is evidence that in general the pre-existing dislocations remain firmly locked during plastic deformation. On the other hand, if precipitate particles or interstitial or vacancy aggregates lock the dislocation line then thermal fluctuations are unlikely to be effective in assisting in the generation of mobile dislocations. Thus k_y should be temperature independent. Most of the evidence on body-centred cubic metals suggests that k_y is temperature independent and consequently yielding is unlikely to occur by the breaking away of locked dislocations. On the contrary, electron microscope evidence shows that dislocation lines are locked by fine precipitates rather than by impurity atmospheres. They also show that above 300 K the dislocation density increases markedly with increasing temperature. This argues that at the higher temperature new dislocations are continually being created in order to maintain the imposed strain rate. As the new dislocations are generated they rapidly become permanently pinned by mobile impurities or by impurity precipitates. Irrespective of the origins of the k_y term, the most important contribution to the yield stress by impurities and radiation induced defects is the general increase in the lattice friction stress. This effect is also responsible for the marked temperature dependence of the yield stress.

7.5 Precipitation hardening in alloys

Alloys which display a decreasing solid solubility with decreasing temperature are frequently strengthened by suitable quench/anneal treatments designed to produce a fine dispersion of precipitate particles. The most effective hardening technique is usually to first solution heat-treat the alloy of composition C at temperature T_q (Fig. 7.11) above the solid solubility limit T_s prior to quenching to a lower temperature. This supersaturated solid solution is usually soft, but becomes harder when precipitation takes place. Precipitation is encouraged by *ageing* for a sufficient length of time at some temperature T_A below T_s. The yield strengths of precipitation-hardened alloys depend critically upon the nature, the size and the separation of the precipitated particles, all of which are controlled by the annealing temperature T_A. The hardening results from the interaction between the impurity precipitate and the dislocation lines. It is observed that the precipitate sizes are finer as the ageing temperature is lowered. A maximum hardening becomes apparent for a critical dispersion of the precipitate particles. However, if ageing is allowed to proceed too far the large number of finely dispersed, small particles is reduced and the precipitate size coarsens. In this over-aged state the alloy becomes softer.

In its simplest form the theory supposes either that the dislocations cut through the precipitate particles (Fig. 7.12) or that they take some convenient path around the obstacles (Fig. 7.12b). When the precipitates are small and finely dispersed in the matrix the glide dislocations can cut through

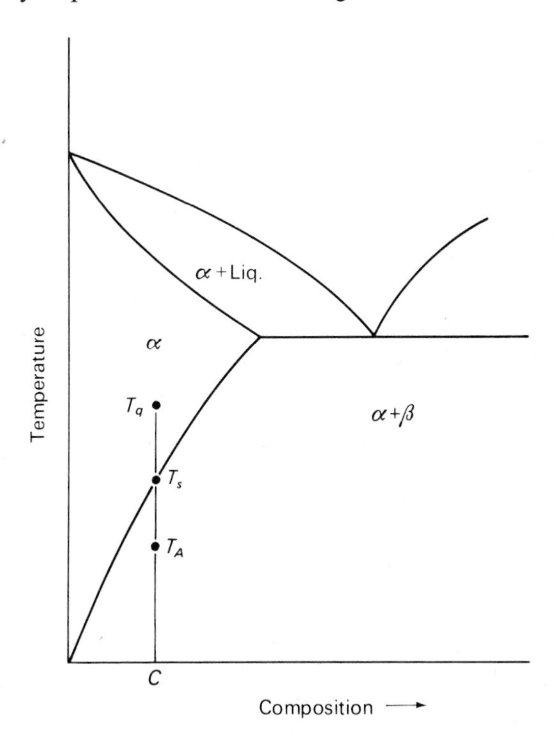

Fig. 7.11 Showing a phase diagram for a possible precipitation hardening alloy.

the particles. Since this process may be thermally activated the yield point is expected to be sensitive to temperature. This is usually the situation obtained in lightly aged alloys. However, the stress level is raised very significantly due to the difficulty in forcing the dislocations through the impurity particles. This is clearly the case for the Al-4·5% Cu alloys as shown by Fig. 7.13. Pure aluminium has a yield stress of only 0·13 kg mm^{-2} while the addition of 4·5% Cu raises the figure to around 3 kg mm^{-2}. Ageing then raises this value still further to over 8 kg mm^{-2}. When the ageing process has resulted in the particle separation being great enough for the dislocations to pass between the particles the yield stress falls to about 1·2 kg mm^{-2}.

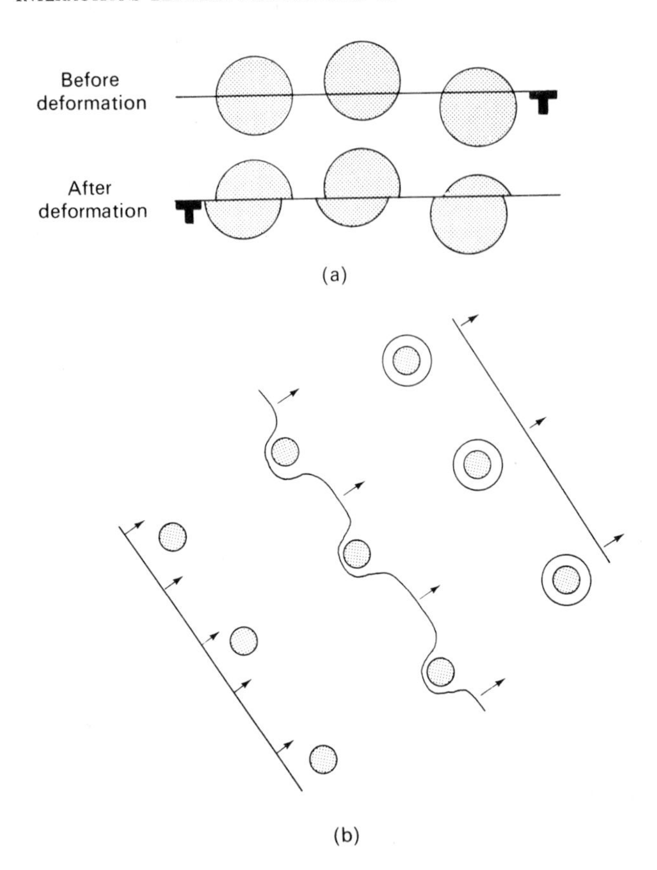

Fig. 7.12 Interaction of dislocation lines with a second phase in an alloy. (a) cutting through closely spaced zones, (b) by-passing precipitates and leaving a residual dislocation loop around each particle.

It is particularly convenient to follow the development of ageing by measuring the hardness of a sample as a function of time, as shown in Fig. 7.14. From this it becomes clear that in Fig. 7.13 curves *B* and *C* respectively correspond to an incompletely aged state and a slightly over-aged state. Figure 7.14 shows that several processes must take place to explain the shape of the hardness/ageing time curves. Recent X-ray and electron microscope studies have greatly assisted in recognizing the different stages in the ageing sequence.[69] As a consequence of these experiments it is evident that at low temperature the first stage in the hardening is due to the development of coherent zones in the matrix. These zones, referred to as Guinier-Preston 1 or GP 1 zones, are typically a few atomic planes thick and about 100 Å along an edge. They are believed to be platelike clusters of copper atoms

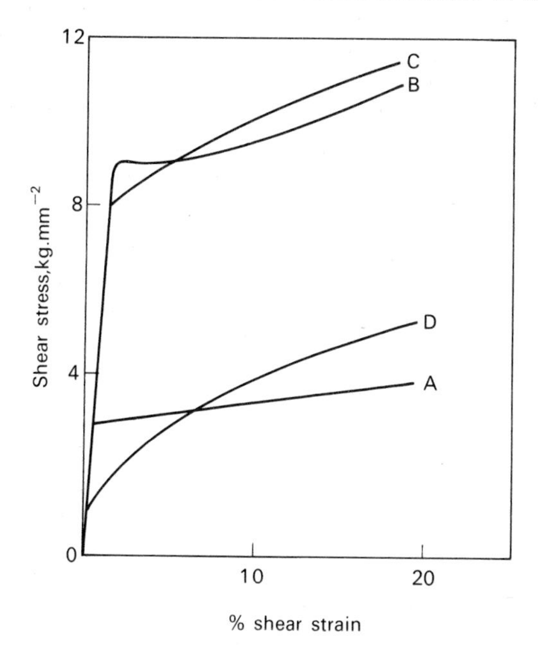

Fig. 7.13 Stress–strain curves of Al-4·5% Cu crystals. A—air cooled, B—aged for 2 days at 130°C, C—aged for 27½ hours at 190°C and D—over-aged at 350°C and slowly cooled. (After GREATHAM and HONEYCOMBE. 1960. *J. Inst. of Met.* **89**, 13.)

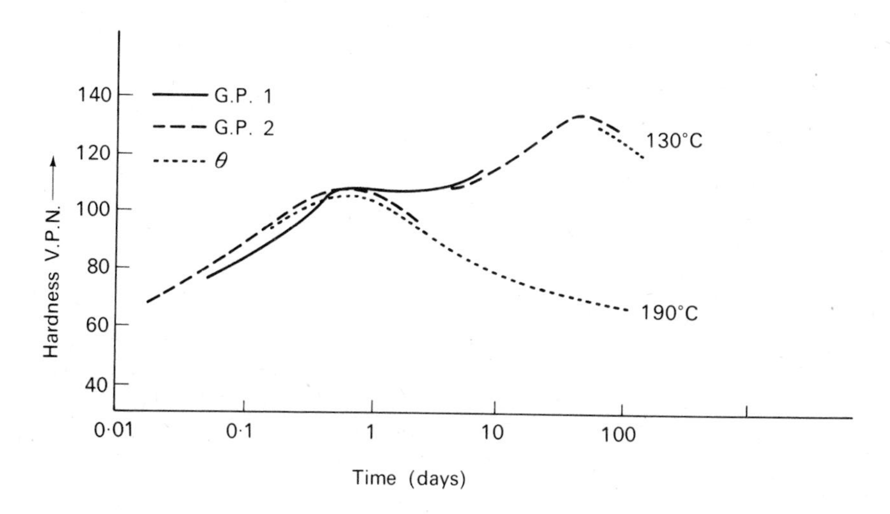

Fig. 7.14 The ageing of aluminium-4·5% Cu alloys as a function of time at constant temperature. (After SILCOCK, HEAL and HARDY, 1953–4. *J. Inst. Met.* **82**, 239.)

segregated on to the {100} planes of the aluminium matrix. There is no interruption in the stacking sequence of the lattice at the zone interface and consequently the zone interface is coherent with the matrix. At long times in the ageing sequence these zones grow until they develop a definite crystal structure, the GP 2 zones, which remain coherent with the matrix since their tetragonal structure fits perfectly with the aluminium unit cell in the *a-a* plane but not perpendicular to this plane. It is evident, therefore, that with both types of zone coherency strains exist. As the precipitates continue to grow to the stage denoted by θ' (when they have a structure modified from that of $CuAl_2$ by the need for a crystallographic relationsip to the matrix they may relieve their coherency strains by the formation of a stable dislocation loop around the precipitate particle. When precipitates have such semicoherent boundaries it is easier for glide dislocations to move through the lattice. The coherency strains disappear entirely with the formation of the equilibrium θ or $CuAl_2$ precipitates. Thus in the Al–Cu precipitation hardening alloys the property changes accompany the sequence GP $1 \rightarrow$ GP $2 \rightarrow \theta' \rightarrow \theta$. The peak hardness corresponds to an optimum distribution of precipitate sizes and coherency strains in the matrix. It is evident in Fig. 7.14 that one or more stages may be omitted when ageing is carried out at higher temperatures.

Precipitation hardening also occurs in numerous other alloy systems, aluminium-based alloys being particularly suited for phase precipitation. The systems Al–Cu, Ag–Al, Al–Mg–Si, Al–Mg–Cu and Al–Mg–Zn have all been extensively investigated. Copper-based alloys containing about 2% Be are also hardenable to such an extent that they produce the strongest and hardest known copper alloys. Nickel and iron are other technologically important age hardening matrices. The possibilities for high temperature strengthening of metallic structures are of course obvious. However, there is a severe problem in maintaining a fine dispersion of precipitates necessary for maximum strength at the high temperature. Thus ordinary precipitation-hardened alloys are not usually useful at high temperature. It is possible to avoid this problem in systems where either (a) precipitate particles have little solubility except near the melting point (e.g. some alloy carbides in steels) or (b) precipitate particles have a structure which matches the parent metal so well that there is no driving force for coarsening of the particles (e.g. Ni_3Al in nickel-based alloys).

In the transition metal carbides the precipitation hardening morphology is different from that discussed above since precipitation occurs preferentially at dislocation nodes. In both titanium carbide and vanadium carbide the presence of small quantities of boron yields impressive increases in strength even at temperatures up to 1700°C as shown in Fig. 7.15. In both cases the increased strength has been associated with the formation of boride precipitates. At least when a high concentration of boron is present the boride precipitates form on the {111} planes. Venables[70] has demonstrated that $TiBr_2$ precipitates in titanium carbide are platelets 10 Å thick situated at

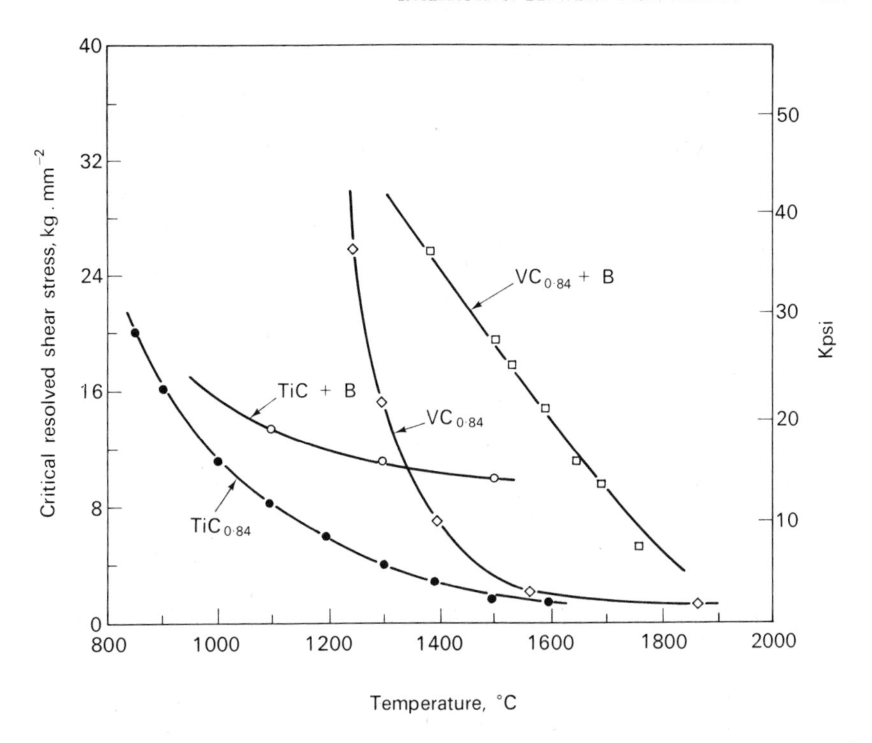

Fig. 7.15 The critical shear stress for slip as a function of temperature in boron-doped TiC and $VC_{0.84}$. (After HOLLOX. 1968/69. *Mat. Sci. Eng.* **3**, 121.)

widely extended dislocation nodes bounded by Shockley partial dislocations (Fig. 7.16). A detailed analysis of the electron diffraction patterns from these precipitates indicates that they exhibit partial coherency with the matrix, since the basal planes of $TiBr_2$ have the same close-packed arrangement as do the {111} planes of titanium carbide. Diffraction contrast experiments on the dislocation nodes reveal that the nucleation sites are three-fold dislocation nodes of the extrinsic type. These observations have significant implications in respect of the high temperature mechanical properties of carbides. Venables[70] has suggested that the precipitate size and density, and hence mechanical strength, may be controlled by varying the triple node density prior to or during doping by mechanical treatment.

7.6 Interaction between dislocations

We have previously discussed (§1.3) how dislocation interactions may result in the formation of simple tilt boundaries. There are two additional

Fig. 7.16 $TiBr_2$-precipitates on dislocation nodes in $VC_{0.84}$. (After VENABLES. 1967. *Phil. Mag.* **16,** 873.)

means by which dislocations may interact:

(i) when parallel dislocation lines meet on different slip planes sessile dislocations may be created,

(ii) dislocations may actually cut through one another thereby becoming jogged or constricted.

These interactions also have important significance for the mechanical properties of solids in view of their capability for degrading the overall dislocation mobility.

The Lomer-Cottrell sessile dislocation

Consider two dissociated dislocations in Fig. 7.17 moving on the $(1\bar{1}1)$ and $(\bar{1}11)$ slip planes. When these dislocations meet they will interact and produce a *stair-rod* dislocation configuration which is sessile. The unit dislocations have Burgers vectors $(a/2)[\bar{1}0\bar{1}]$ and $(a/2)[0\bar{1}1]$ on the slip planes $(1\bar{1}1)$ and $(\bar{1}11)$ respectively. From the reference tetrahedron (see Appendix to Chapter 1) we can represent the dissociation of these dislocations according to the reactions

$$DC \rightarrow D\beta + \beta C$$

and

$$CB \rightarrow C\delta + \delta B$$

Thus the dislocation configuration on each plane consists of a pair of partial dislocations separated by a stacking fault. When the leading partial dislocations meet along the intersection of the slip planes they combine according to

$$\beta C + C\delta \rightarrow \beta\delta$$

Thus the total reaction is written

$$DC + CB \rightarrow B\beta + \beta\delta + \delta B$$

i.e.

$$\frac{a}{2}[10\bar{1}] + \frac{a}{2}[0\bar{1}1] \rightarrow \frac{a}{6}[\bar{1}\bar{1}\bar{2}] + \frac{a}{6}[110] + \frac{a}{6}[\bar{1}\bar{1}2]$$

Since the dislocation energy $E \propto \boldsymbol{b}^2$ there is a lowering of energy resulting from this interaction according to

$$\frac{a^2}{2} + \frac{a^2}{2} < \frac{a^2}{6} + \frac{a^2}{3} + \frac{a^2}{6}$$

Consequently the interaction is energetically favoured. The *stair-rod* or Lomer-Cottrell dislocation $\beta\delta$ is joined to both Shockley partial dislocations by the two regions of stacking fault. Notice also that the dislocation $\beta\delta$ is sessile since the Burgers vector does not lie in either of the planes $(1\bar{1}1)$ and $(\bar{1}11)$ which contain the stacking fault.

Strain hardening in crystals results from new barriers to dislocation movement being introduced during deformation. The Lomer-Cottrell dislocation has obvious potential in this capacity, and will present a formidable obstacle to slip on both the intersecting slip planes. This is especially true when the stacking fault energy is low in view of the extent of the wedge-shaped bands of stacking fault. When the stacking fault energy is high the separation between the partials is not as great and such barriers are easily broken down by the action of a dislocation pile-up.

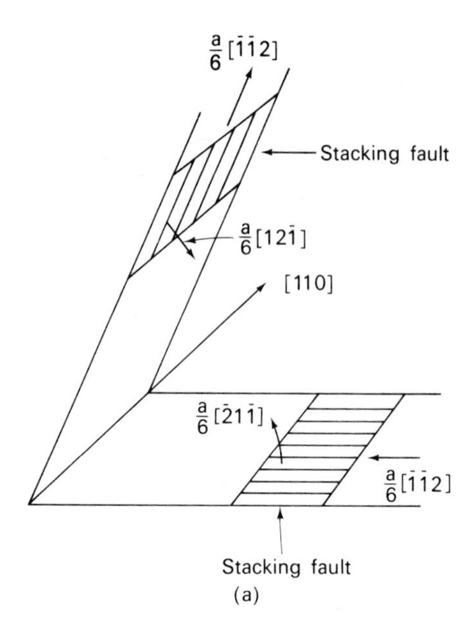

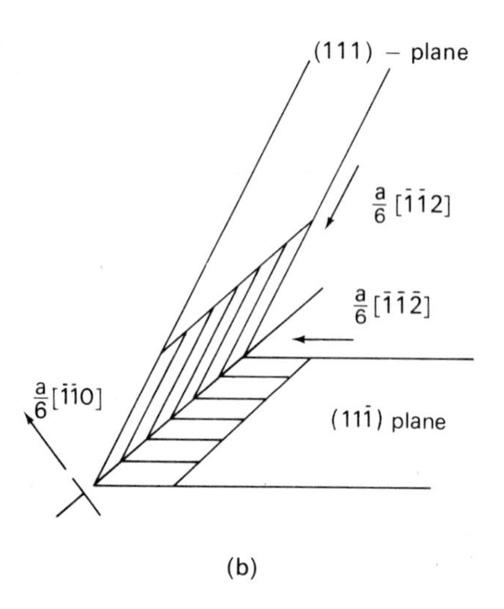

Fig. 7.17 Formation of Lomer-Cottrell sessile dislocation on intersecting {111} plane. (a) Shows the two dissociated dislocations prior to the formation of the stair-rod dislocations in (b).

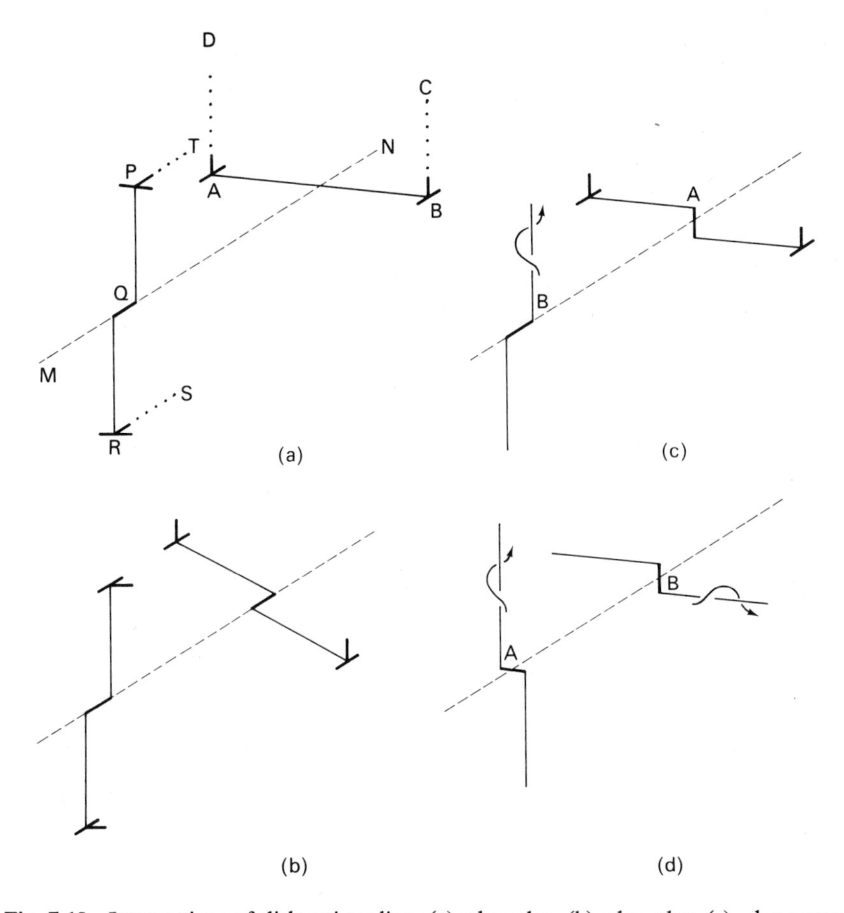

Fig. 7.18 Intersections of dislocations lines (a) edge-edge, (b) edge-edge, (c) edge-screw, (d) screw-screw. (After COTTRELL. 1953. *Dislocation and Plastic Flow in Crystals.* Oxford University Press, Oxford.)

Intersecting dislocations

When dislocation lines pass through one another they may result in jogs being produced. Examples of this are shown in Fig. 7.18. Consider first the intersection of two edge dislocations (Fig. 7.18a). Dislocation PR is jogged at Q by a dislocation AB cutting through it. The relative displacements of the crystal above and below the slip plane (i.e. the length of the jog) are given by the Burgers vector of dislocation PR. Similarly in Fig. 7.18c the edge dislocation A is jogged when it intersects the screw dislocation B: the screw dislocation is also jogged in this process. In much the same way we can see that

(a) two edge dislocations with parallel Burgers vectors both contain jogs after intersecting one another (Fig. 7.18b)

(b) two screw dislocations both contain jogs after intersecting one another (Fig. 7.18d).

These jogs may exert a profound influence on the subsequent motion of dislocations. In Fig. 7.18a the jog J has a Burgers vector which is perpendicular to J and it is therefore an edge dislocation. Thus this jog will glide along with dislocation, since its Burgers vector is all the time in the slip plane. However, when two edge dislocations with parallel Burgers vector intersect the resulting jogs are in the screw orientation and lie in the slip plane. Thus jogs in pure edge dislocations do not impede the subsequent motion of these dislocations. On the other hand all jogs in screw dislocations are short lengths of edge dislocations. Consequently such jogs are constrained to move *along* the dislocation line since this is the direction of its Burgers vector. Thus if the screw dislocation glides forward in its normal manner, the jog must trail behind it since it cannot glide in this direction. Thus the screw dislocation is pinned by the jog. At high stresses it is possible that the screw dislocation can be induced to move, trailing the jog behind it. This non-conservative motion leaves a trail of interstitials or vacancies in the wake of the jog, depending upon the sign of the dislocation and the direction of motion. With dissociated dislocations the jogging process is particularly difficult since in crossing the stacking fault the moving dislocation would tend to produce a complex fault in the plane of intersection. Even if the partial dislocations coalesce to form a constriction a large expenditure of energy is required, particularly if the stacking fault is wide.

Some further consequence of dislocation dissociation

There is considerable experimental evidence to support the notion developed in Chapter 1 that in materials with low stacking fault energy dislocations are dissociated into Shockley partial dislocations. Stacking faults have now been observed in thin films of zinc, copper, silver, gold, graphite, silicon and many alloys. Frequently, dissociated dislocations take the form of dislocation networks which develop *nodes* as a consequence of the interactions between the extended dislocations. When three or more dislocations meet at a point (or node) it is geometrically necessary that $\sum_{1}^{n} \boldsymbol{b}_i = 0$, where the $\boldsymbol{b}_i$ are the Burgers vectors of the dislocations in the node. In materials with low stacking fault energy the dislocation interactions give a series of extended and contracted nodes depicted schematically in Fig. 7.19. The development of these networks from hexagonal dislocation networks is

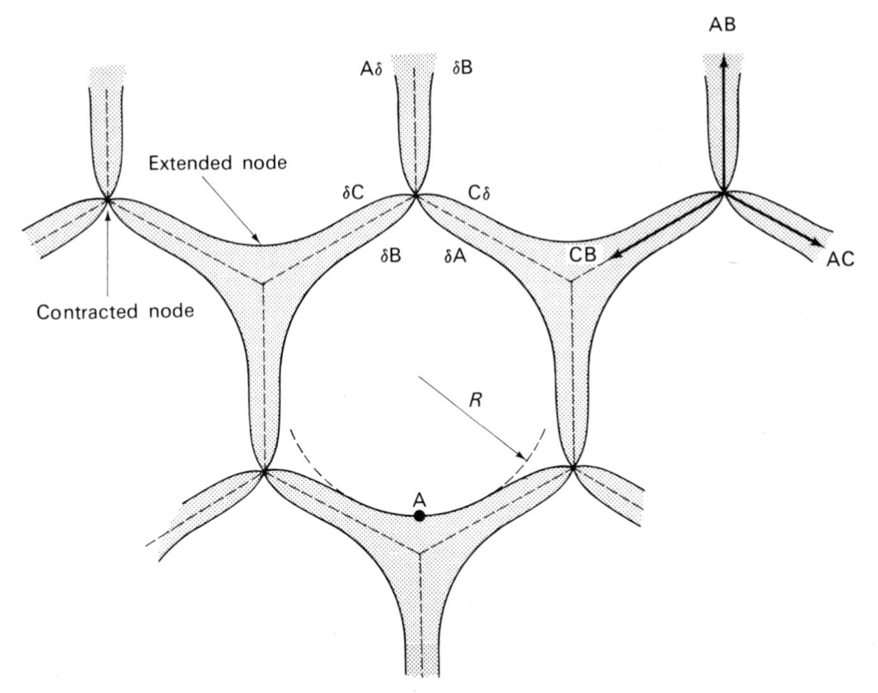

Fig. 7.19 Schematic representation of a hexagonal network of extended and contracted nodes in a face-centred cubic lattice.

evident. The properties of these nodes are easily understood in terms of the dislocation terminology developed from the Thompson tetrahedron.

The nodes shown in Fig. 7.19 clearly result from the interactions between three dislocations. Consider that all three dislocations lie on the same plane e.g. the (111) plane. Thus the nodal condition requires that the three dislocations have Burgers vectors which are some cyclic arrangement of the three unit dislocations AB, BC and CA. When the dislocation with Burgers vector AB on the (111) plane intersects a dislocation on the $(11\bar{1})$ plane with Burgers vectors BC the resulting arrangement is that shown in Fig. 7.19. Initially the two dislocations dissociate into partials according to

$$AB \rightarrow A\delta + \delta B$$
$$CB \rightarrow C\alpha + \alpha B$$

The dislocations constrict at the intersection and CB then dissociates into the (111) plane according to

$$CB \rightarrow C\delta + \delta B$$

Thus the partial dislocations at the nodes all lie on the same slip plane. Consequently these nodes are fully mobile in the (111) plane. There is an obvious similarity between the schematic representation in Fig. 7.19 and the electron micrograph in Fig. 7.16 of dislocation nodes in titanium carbide.

Whelan[71] has shown that the equilibrium radius of curvature R of the extended node may be used to estimate the stacking fault energy. The partial dislocation at A is in equilibrium under the force per unit length tending to straighten the dislocation and the elastic forces due to the stacking fault. If the line energy is T and γ the stacking fault energy then equating the forces we have

$$\gamma = T/R = \alpha\mu b^2/R$$

The equation is best used in materials with low stacking fault energy. High purity is, however, essential since Fig. 7.16 shows that trace impurities may result in extensive dissociation on account of the impurity segregation effects. Some typical values of the stacking fault energy measured using the node analysis are given in Table 7.1.

Table 7.1 Stacking fault energies of some simple metals

Material	Stacking fault energy J cm^{-2}	Method
Stainless steel	<10	node analysis
Silver	25	node analysis
Silver + 25% Zn	3	node analysis
Gold	45	node analysis
Copper	70	node analysis
Copper + 25% Zn	7	node analysis
Aluminium	135	Loop analysis
Magnesium	125	Loop analysis
Zinc	140	Loop analysis
Nickel	225	Stage III work hardening

GENERAL REFERENCES

COTTRELL, A. H. 1953. *Dislocations and Plastic Flow in Crystals.* Oxford University Press, Oxford.

SMALLMAN, R. E. 1961. *Modern Physical Metallurgy.* Butterworths, London.

THOMAS, G. and WASHBURN, J. (Eds) 1963. *Electron Microscopy and the Strength of Metals.* Interscience, New York and London.

HONEYCOMBE, R. W. K. 1968. *The Plastic Deformation of Metals.* Arnold, London.

Bibliography

1. Mott, N. F. and Jones, H. 1936. *Theory of the Properties of Metals and Alloys.* Oxford University Press, Oxford.
2. Cottrell, A. H. 1953. *Dislocation and Plastic Flow in Crystals.* Oxford University Press, Oxford.
3. Burgers, J. M. 1939. *Proc. Kon. Ned. Akad. Wet.* **49,** 293 and 378.
4. Read, W. T. 1953. *Dislocation in Crystals.* McGraw-Hill, New York.
5. Parker, E. R. and Washburn, J. 1952. *J. Metals.* **4,** 1076.
6. Lomer, W. M. and Nye, J. F. 1952. *Proc. Roy. Soc.* **A212,** 576.
7. Mott, N. F. 1948. *Proc. Phys. Soc.* **60,** 391.
8. Chadderton, L. T. 1965. *Radiation Damage in Crystals.* Methuen, London.
9. Kinchin, G. S. and Pease, R. S. 1955. *Rep. Prog. Phys.* **18,** 1.
10. Corbett, J. W. 1966. *Electron Irradiation in Semiconductors.* Academic Press, New York.
11. Ziman, J. M. 1960. *Electrons and Phonons.* Clarendon Press, Oxford.
12. Wilmhurst, T. H. 1968. *Electron Spin Resonance Spectrometers.* Hilger, London.
13. Thomas, G. 1962. *Transmission Electron Microscopy of Metals.* John Wiley, London.
14. Lidiard, A. B. 1957. *Handbuch der Physik.* **20,** 246.
15. Seitz, F. 1946. *Rev. Mod. Phys.* **18,** 384; and 1954. ibid. **26,** 7.
16. Barr, L. W. and Lidiard, A. B. 1971. *Physical Chemistry—An Advanced Treatise,* **10** (Ed. Eyring, Ac. Press, N.Y.)
17. Grundig, H. 1965. *Z. Physik.* **182,** 477.
18. Barr, L. W. and Dawson, D. K. 1968. Abstract 13 *Colour Centres in Alkali Halides.* International Symposium, Rome; and 1967. *Proc. Brit. Ceram. Soc.* **9,** 171.
19. Franklin, A. D. (Ed.) 1967. *Calculations of the Properties of Vacancies and Interstitials.* N.B.S. Special Publication No. 287 and references therein. Washington D.C.
20. Jost, W. 1933. *J. Chem. Phys.* **1,** 466.
21. Schulman, J. H. and Compton, W. D. 1962. *Colour Centres in Solids.* Pergamon Press, Oxford.
22. De Boer, J. H. 1937. *Rec. Trav. Chim. Pays-Bas.* **56,** 301.
23. H. Seidel and H. C. Wolf in *The Physics of Colour Centres,* Ed. Fowler, 1968, Academic Press, New York.
24. Fowler, W. B. Ed. 1968. *The Physics of Colour Centres.* Academic Press, New York.
25. Lax, M. 1952. *J. Chem. Phys.* **20,** 1752.
26. Swank, R. K. and Brown, F. C. 1963. *Phys. Rev.* **130,** 34.
27. Luty, F. 1968. *The Physics of Colour Centres.* Ed. Fowler, Academic Press, New York.
28. Condon, E. U. and Shortley, G. H. 1935. *Theory of Atomic Spectra.* Cambridge University Press, London.
29. Ivey, H. 1947. *Phys. Rev.* **72,** 341.
30. Ueta, M. 1952. *J. Phys. Soc. Jap.* **7,** 107.
31. Faraday, B., Rabin, H. and Compton, W. D. 1961. *Phys. Rev. Lett.* **7,** 57.
32. Silsbee, R. H. 1965. *Phys. Rev.* **138,** A180.
33. Krupka, D. C. and Silsbee, R. H. 1966. *Phys. Rev.* **152,** 816.

34. HUGHES, A. E. and RUNCIMAN, W. A. 1965. *Proc. Roy. Soc.* **86,** 615.
35. FITCHEN, D. B. 1968. in *The Physics of Colour Centres.* Ed. Fowler, Academic Press, New York.
36. SEIDEL, H., SCHWOERER, M. and SCHMID, D. 1965. *Z. Physik.* **182,** 398.
37. HUGHES, A. E. 1966. *Proc. Roy. Soc.* **87,** 535 and **88,** 449.
38. CASTNER, T. G., KANZIG, W. and WOODRUFF, T. O. 1958. *Nuovo Cimento Supp.* (10), see for example **7,** 612 and references therein.
39. SLICHTER, C. P. 1963. *Magnetic Resonance.* Harper and Row, New York.
40. BROWN, F. C. 1966. *The Physics of Solids.* Benjamin, New York.
41. WANNIER, G. 1937. *Phys. Rev.* **52,** 191.
42. DRESSELHAUS, G., KIP, A. and KITTELL, C. 1955. *Phys. Rev.* **98,** 368.
43. SEEGER, A. and SWANSON, M. L., 1968, *International Symposium on Lattice Defects in Semiconductors.* Ed. Hasiguti, Tokyo University Press.
44. SEEGER, A. and CHIK, K. P. 1968. *Phys. Stal. Sol.* **29,** 455.
45. For a discussion of Fick's Laws see COTTRELL, A. H. 1953. *Theoretical and Structural Metallurgy.* Arnold, London.
46. WATKINS, G. D. 1968. *Radiation Effects in Semiconductors.* Plenum Press, New York p. 67 and references therein.
47. WATKINS, G. D. and CORBETT, J. W. 1964. *Phys. Rev.* **134,** A1359.
48. YAMAGUCHI, T. 1962. *J. Phys. Soc.* Jap. **17,** 1359.
49. WHAN, R. E. 1966. *J. App. Phys.* **37,** 3378.
50. ALEXANDER, H. and HAASEN, P. 1968. *Sol. St. Phys.* **22,** 27.
51. SIMMONS, R. O. and BALLUFFI, R. W. 1960. *Phys. Rev.* **117,** 52 and 62.
52. BALL, A. 1969. *Phil. Mag.* **20,** 113.
53. VENABLES, J. D., KAHN, D. and LYE, R. G. 1968. *Phil. Mag.* **18,** 177.
54. KINO, T. and KOEHLER, J. S. 1967. *Phys. Rev.* **162,** 632.
55. DOBSON, P S. and SMALLMAN, R. E. 1966. *Proc. Roy. Soc.* **A293,** 423.
56. DOBSON, P. S., GOODHEW, P. J. and SMALLMAN, R. E. 1967. *Phil. Mag.* **16,** 9.
57. CORBETT, J. W., SMITH, R. B. and WALKER, R. M. 1959. *Phys. Rev.* **114,** 1452 and 1460.
58. LOMER, J. N. and TAYLOR, M. R. 1969. *Phil. Mag.* **19,** 437.
59. HOLLOX, G. E. 1968/69. *Mater. Sci. Eng.* **3,** 121.
60. GILMAN, J. J. and JOHNSTON, W. G. 1962. *Solid State Physics,* **13,** 147. Academic Press, New York.
61. COTTRELL, A. H. and BILBY, B. A. 1949. *Proc. Phys.* **62,** 49.
62. BULLOUGH, R. and NEWMAN, R. C. 1963. *Progress in Semiconductors.* **7,** 100.
63. FLEISCHER, R. L. 1961. *Acta Met.* **9,** 996; 1962. *Acta Met.* **10,** 835; *J. App. Phys.* **33,** 3504; 1963. *Acta Met.* **11,** 203.
64. MOTT, N. F. 1937. *Proc. Phys. Soc.* **49,** 258.
65. COTTRELL, A. H., HUNTER, S. C. and NABARRO, F. R. N. 1953. *Phil. Mag.* **44,** 1064.
66. SUZUKI H., 1957. *Dislocations and Mechanical Properties of Crystals.* Eds. Fisher et al. Wiley, New York.
67. COTTRELL, A. H. 1963. in "The relation between Structure and Mechanical Properties of Metals" H.M.S.O. London.
68 PETCH, N. J. 1953. *J. Iron Steel Inst.* **173,** 25.
69. KELLY, A. and NICHOLSON, R. B. 1963. *Prog. Mat. Science.* **10,** 151.
70. VENABLES, J. D. 1967. *Phil. Mag.* **16,** 873.
71. WHELAN, M. J. 1959. *Proc. Roy. Soc.* **A249,** 114.

Indexes

Author Index

Subject Index